Chapman & Hall/CRC Mathematical and Computational Biology Series

Optimal Control Applied to Biological Models

CHAPMAN & HALL/CRC
Mathematical and Computational Biology Series

Aims and scope:
This series aims to capture new developments and summarize what is known over the whole spectrum of mathematical and computational biology and medicine. It seeks to encourage the integration of mathematical, statistical and computational methods into biology by publishing a broad range of textbooks, reference works and handbooks. The titles included in the series are meant to appeal to students, researchers and professionals in the mathematical, statistical and computational sciences, fundamental biology and bioengineering, as well as interdisciplinary researchers involved in the field. The inclusion of concrete examples and applications, and programming techniques and examples, is highly encouraged.

Series Editors
Alison M. Etheridge
Department of Statistics
University of Oxford

Louis J. Gross
Department of Ecology and Evolutionary Biology
University of Tennessee

Suzanne Lenhart
Department of Mathematics
University of Tennessee

Philip K. Maini
Mathematical Institute
University of Oxford

Shoba Ranganathan
Research Institute of Biotechnology
Macquarie University

Hershel M. Safer
Weizmann Institute of Science
Bioinformatics & Bio Computing

Eberhard O. Voit
The Wallace H. Couter Department of Biomedical Engineering
Georgia Tech and Emory University

Proposals for the series should be submitted to one of the series editors above or directly to:
CRC Press, Taylor & Francis Group
24-25 Blades Court
Deodar Road
London SW15 2NU
UK

Published Titles

Cancer Modelling and Simulation
Luigi Preziosi

Computational Biology: A Statistical Mechanics Perspective
Ralf Blossey

Computational Neuroscience: A Comprehensive Approach
Jianfeng Feng

Data Analysis Tools for DNA Microarrays
Sorin Draghici

Differential Equations and Mathematical Biology
D.S. Jones and B.D. Sleeman

Exactly Solvable Models of Biological Invasion
Sergei V. Petrovskii and Bai-Lian Li

Introduction to Bioinformatics
Anna Tramontano

An Introduction to Systems Biology: Design Principles of Biological Circuits
Uri Alon

Knowledge Discovery in Proteomics
Igor Jurisica and Dennis Wigle

Modeling and Simulation of Capsules and Biological Cells
C. Pozrikidis

Niche Modeling: Predictions from Statistical Distributions
David Stockwell

Normal Mode Analysis: Theory and Applications to Biological and Chemical Systems
Qiang Cui and Ivet Bahar

Optimal Control Applied to Biological Models
Suzanne Lenhart and John T. Workman

Stochastic Modelling for Systems Biology
Darren J. Wilkinson

The Ten Most Wanted Solutions in Protein Bioinformatics
Anna Tramontano

Chapman & Hall/CRC Mathematical and Computational Biology Series

Optimal Control Applied to Biological Models

Suzanne Lenhart
John T. Workman

Chapman & Hall/CRC
Taylor & Francis Group
Boca Raton   London   New York

Chapman & Hall/CRC is an imprint of the
Taylor & Francis Group, an **informa** business

Chapman & Hall/CRC
Taylor & Francis Group
6000 Broken Sound Parkway NW, Suite 300
Boca Raton, FL 33487-2742

© 2007 by Taylor & Francis Group, LLC
Chapman & Hall/CRC is an imprint of Taylor & Francis Group, an Informa business

No claim to original U.S. Government works
Printed in the United States of America on acid-free paper
10 9 8 7 6 5 4 3 2 1

International Standard Book Number-10: 1-58488-640-4 (Hardcover)
International Standard Book Number-13: 978-1-58488-640-2 (Hardcover)

This book contains information obtained from authentic and highly regarded sources. Reprinted material is quoted with permission, and sources are indicated. A wide variety of references are listed. Reasonable efforts have been made to publish reliable data and information, but the author and the publisher cannot assume responsibility for the validity of all materials or for the consequences of their use.

No part of this book may be reprinted, reproduced, transmitted, or utilized in any form by any electronic, mechanical, or other means, now known or hereafter invented, including photocopying, microfilming, and recording, or in any information storage or retrieval system, without written permission from the publishers.

For permission to photocopy or use material electronically from this work, please access www.copyright.com (http://www.copyright.com/) or contact the Copyright Clearance Center, Inc. (CCC) 222 Rosewood Drive, Danvers, MA 01923, 978-750-8400. CCC is a not-for-profit organization that provides licenses and registration for a variety of users. For organizations that have been granted a photocopy license by the CCC, a separate system of payment has been arranged.

Trademark Notice: Product or corporate names may be trademarks or registered trademarks, and are used only for identification and explanation without intent to infringe.

Library of Congress Cataloging-in-Publication Data

Lenhart, Suzanne.
   Optimal control applied to biological models / Suzanne Lenhart and John T. Workman.
      p. cm. -- (Mathematical and computational biology ; 15)
      Includes bibliographical references (p.   ) and index.
   ISBN-13: 978-1-58488-640-2 (alk. paper)
   ISBN-10: 1-58488-640-4 (alk. paper)
   1. Biological models--Textbooks. 2. Mathematical optimization--Textbooks. 3. Control theory--Textbooks. I. Workman, John T. II. Title. III. Series.

QH324.8.L46 2007
570.15'1--dc22
                                                                             2007000973

Visit the Taylor & Francis Web site at
http://www.taylorandfrancis.com

and the CRC Press Web site at
http://www.crcpress.com

Contents

Preface	**xi**
1 Basic Optimal Control Problems	**1**
1.1 Preliminaries	4
1.2 The Basic Problem and Necessary Conditions	7
1.3 Pontryagin's Maximum Principle	12
1.4 Exercises	18
2 Existence and Other Solution Properties	**21**
2.1 Existence and Uniqueness Results	23
2.2 Interpretation of the Adjoint	26
2.3 Principle of Optimality	28
2.4 The Hamiltonian and Autonomous Problems	31
2.5 Exercises	35
3 State Conditions at the Final Time	**37**
3.1 Payoff Terms	37
3.2 States with Fixed Endpoints	41
3.3 Exercises	46
4 Forward-Backward Sweep Method	**49**
5 Lab 1: Introductory Example	**57**
6 Lab 2: Mold and Fungicide	**63**
7 Lab 3: Bacteria	**67**
8 Bounded Controls	**71**
8.1 Necessary Conditions	73
8.2 Numerical Solutions	81
8.3 Exercises	83
9 Lab 4: Bounded Case	**85**
10 Lab 5: Cancer	**89**
11 Lab 6: Fish Harvesting	**93**

12 Optimal Control of Several Variables — 97
- 12.1 Necessary Conditions — 97
- 12.2 Linear Quadratic Regulator Problems — 104
- 12.3 Higher Order Differential Equations — 107
- 12.4 Isoperimetric Constraints — 108
- 12.5 Numerical Solutions — 112
- 12.6 Exercises — 113

13 Lab 7: Epidemic Model — 117

14 Lab 8: HIV Treatment — 123

15 Lab 9: Bear Populations — 129

16 Lab 10: Glucose Model — 135

17 Linear Dependence on the Control — 139
- 17.1 Bang-Bang Controls — 139
- 17.2 Singular Controls — 143
- 17.3 Exercises — 151

18 Lab 11: Timber Harvesting — 153

19 Lab 12: Bioreactor — 157

20 Free Terminal Time Problems — 163
- 20.1 Necessary Conditions — 163
- 20.2 Time Optimal Control — 168
- 20.3 Exercises — 173

21 Adapted Forward-Backward Sweep — 175
- 21.1 Secant Method — 175
- 21.2 One State with Fixed Endpoints — 177
- 21.3 Nonlinear Payoff Terms — 182
- 21.4 Free Terminal Time — 183
- 21.5 Multiple Shots — 184
- 21.6 Exercises — 187

22 Lab 13: Predator-Prey Model — 189

23 Discrete Time Models — 193
- 23.1 Necessary Conditions — 193
- 23.2 Systems Case — 199
- 23.3 Exercises — 202

24 Lab 14: Invasive Plant Species — 205

25 Partial Differential Equation Models **211**
   25.1 Existence of an Optimal Control 212
   25.2 Sensitivities and Necessary Conditions 213
   25.3 Uniqueness of the Optimal Control 215
   25.4 Numerical Solutions 215
   25.5 Harvesting Example 216
   25.6 Beaver Example 220
   25.7 Predator-Prey Example 223
   25.8 Identification Example 228
   25.9 Controlling Boundary Terms 231
   25.10 Exercises 234

26 Other Approaches and Extensions **237**

References **245**

Index **259**

Preface

Consider a system in some application, where the dynamics are captured by a model, whether it be ordinary differential equations (ODEs), partial differential equations (PDEs), or discrete difference equations. Suppose also this system has a variable, or variables, which can be controlled from the outside. The question which naturally arises is how exactly to control this element in order to produce the "best" outcome, as measured by some predetermined goal or goals. The mathematical theory behind answering these questions, often called optimal control theory or dynamic optimization, has found application in a myriad of fields, from the biological sciences, to economics, to business and management, to physics and engineering.

The goal of this text is two fold. First, we wish to present the reader with an introductory, but thorough, development of the mathematical aspects of optimal control theory. This is done in a "graded" way, as the most basic problem, with a continuous time ODE, is examined in Chapter 1, and increasingly more complicated problems are handled as the book progresses. This includes variations of the initial conditions, imposed bounds on the control, multiple states and controls, linear dependence on the control, and free terminal time. Optimal control of discrete systems and optimal control of partial differential equations are also introduced.

The second goal is to give the reader an insight into application of optimal control theory to biological models. Several different kinds of applications are presented here, including disease models of immunology and epidemic types, management decisions in harvesting and resource allocation models, and more. These are presented in the interactive "lab" sections, which we feel is a novel feature of this text. The MATLAB codes on which the labs are based are included, in addition to a user-friendly interface, which will allow everyone, even those with no prior MATLAB knowledge, to access them. The underlying numerical methods are also developed in the text.

This book is designed for use as a textbook for advanced undergraduate or beginning graduate students. It would be suitable for a one-semester course. It can also be used by anyone who wants to learn optimal control theory for application to specific models. Mathematically, only a basic knowledge of multi-variable calculus and simple ordinary differential equations is needed for the bulk of the text. Some prior knowledge of PDEs is required for the (optional) chapter on this subject. The reader should also be familiar with mathematical models and how they are used. This book is not intended as a course in mathematical modeling.

Each so-called "theory" chapter has several fully-worked examples and ends with a group of exercises. There are also, throughout the book, more open-ended and thought provoking questions dealing with specific models or applications. The reader is advised to take advantage of both kinds of exercises.

We view this book as an introduction; the last chapter provides some information about more advanced topics. We have also tried to provide references for further reading. This includes papers and other texts where one can find additional information on theoretical, numerical, or biological questions.

We recall the impact of the tools of dynamic programming on the field of behavioral ecology resulting from the work of Clark, Mangel, Houston, and McNamara [34, 86, 136]. We hope that some biologists will consider using the tools introduced here for new applications.

The idea for this book came while working on materials for the short course *Optimal Control Theory in Application to Biology*. This short course, sponsored by the National Institutes of Health, took place at the University of Tennessee in the summer of 2003.

The authors would like to take this opportunity to thank several people who have helped immensely during the preparation of this book: Chuck Collins, for his numerical guidance and all our chats; Mike Saum and Hem Raj Joshi, for their technical expertise; Elsa Schaefer and Lou Gross, for their helpful suggestions; and Peter Andreae, Wandi Ding, Renee Fister, Elizabeth Martin, Vladimir Protopopescu, and Raj Soni for their help in various ways. We would also like to acknowledge the many authors, on whose work several of the examples and labs are based.

To download the MATLAB m-files needed for the labs, go to `www.math.utk.edu/~lenhart/mfiles.d`. Send any questions or comments about this book to `lenhart@math.utk.edu`.

Suzanne Lenhart
University of Tennessee
Oak Ridge National Laboratory

and

John T. Workman
Cornell University

Chapter 1

Basic Optimal Control Problems

We present a motivating idea of optimal control theory in a classic application from King and Roughgarden [104] on allocation between vegetative and reproductive growth for annual plants. This plant growth model formulated by Cohen [36] divides the plant into two parts: the vegetative part, consisting of leaves, stems, and roots, and the reproductive part. The products of photosynthesis (growth) are partitioned into these parts, and the rate of photosynthesis is assumed to be proportional to the weight of the vegetative part. Let $x_1(t)$ be the weight of the vegetative part at time t and $x_2(t)$ the weight of the reproductive part. Consider the following ordinary differential equation model:

$$x_1'(t) = u(t)x_1(t),$$
$$x_2'(t) = (1 - u(t))x_2(t),$$
$$0 \leq u(t) \leq 1,$$
$$x_1(0) > 0,\ x_2(0) \geq 0,$$

where the function $u(t)$ is the fraction of the photosynthate partitioned to vegetative growth. The natural evolution of the plant should encourage maximal growth of the reproductive part in order to ensure effective reproduction. Therefore, the goal is to find a partitioning pattern control $u(t)$ which maximizes the functional

$$\int_0^T \ln(x_2(t))\, dt.$$

The maximum season length is the upper bound T on the time interval, and it is assumed that all season lengths from zero to a fixed maximum have equal probability of occurrence. The natural logarithm appears here because it is believed the evolution of the plant favors reproduction in a nonlinear way.

This type of problem is called an optimal control problem, because we are charged with finding an optimal control, i.e., a control which optimizes some objective functional. We would say that this problem has two states, x_1 and x_2, and one control, u. King and Roughgarden used optimal control theory to solve this problem. Figure 1.1 gives an example of an optimal control for the case $T = 5$.

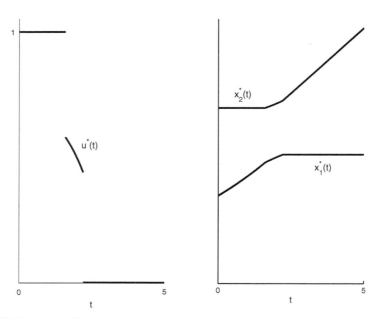

FIGURE 1.1: The optimal photosynthate u^* is shown on the left, and the optimal vegetative and reproductive weights, x_1^* and x_2^*, are on the right.

Analyzing such a problem with a variety of T values can give interesting conclusions. Their analysis leads to the prediction that annual plants experiencing variable length seasons will exhibit graded strategies, with vegetative and reproductive growth occurring simultaneously during part of the life cycle. In other words, the plant will use all of its photosynthate for vegetative growth and later will split it into some vegetative and some reproductive growth.

The goal of this book is to give an introduction to optimal control theory as applied to biological models. Using optimal control theory, one can adjust controls in a system to achieve a goal, where the underlying system can include:

- Ordinary differential equations

- Partial differential equations

- Discrete equations

- Stochastic differential equations

Basic Optimal Control Problems

- Integro-difference equations

- Combination of discrete and continuous systems.

Our primary focus in this text is optimal control theory of ordinary differential equations with time as the underlying variable. Optimal control of discrete equations and PDEs is discussed in Chapters 23 and 25, respectively. For other types of systems, see [10, 11, 128, 129, 182].

Optimal control theory is a powerful mathematical tool that can be used to make decisions involving complex biological situations. For example, what percentage of the population should be vaccinated as time evolves in a given epidemic model to minimize the number of infected and the cost of implementing the vaccination strategy? The desired outcome, or goal, depends on the particular situation. Many times, the problem will include tradeoffs between two competing factors. For another example, consider minimizing a certain harmful virus population while keeping the level of the toxic drug administered low. In such a case, we could model the levels of virus and drug as functions of time appearing together in a system of ordinary differential equations.

The behavior of the underlying dynamical system is described by a **state** variable(s). We assume that there is a way to steer the state by acting upon it with a suitable **control** function(s). The control enters the system of ordinary differential equations and affects the dynamics of the state system. The goal is to adjust the control in order to maximize (or minimize) a given **objective functional**. A functional, for this text, refers to a map from a certain set of functions to the real numbers (an integral, for example). Often, this functional will balance judiciously the desired goal with the required *cost* to reach it. Here, the *cost* may not always represent money but may include side effects or damages caused by the control. In general, the objective functional depends on one or more of the state and the control variables. Frequently the objective functional is given by an integral of the state and/or control variables. Other types of functionals will be considered as well.

Many applications have several state variables and multiple control variables. The plant problem above has two state variables and one control variable, and is a bit unusual in that the objective functional does not depend on the control. Note that the control variables have imposed bounds of 0 and 1 and that the system and objective functional depend on the control u in a linear way. Problems without control constraints (bounds) are usually easier than those with bounds. Also, problems linear in the control are sometimes trickier than those with a reasonable nonlinearity in the control dependence.

We will treat all these wrinkles, and more, in this book. First, we will concentrate on the case of one control and one state, in which the controls do not have any constraints on them. We will also initially focus on problems in which the control enters the problem in a simple nonlinear way, mostly quadratic.

1.1 Preliminaries

Before beginning, we establish some definitions and concepts from analysis and advanced calculus used throughout the book. It is also advantageous to quickly review a few fundamental results. Wade [177] is an excellent source for these and other basic analytical concepts. Biological terminology will be presented as needed. For some background on models from an undergraduate viewpoint, see the book by Mooney and Swift [146]. Mathematical biology modelling for undergraduates (or graduate students totally new to this topic) is covered in the classic book by Edelstein or the book by Jones and Sleeman [53, 89]. For a beginning graduate student viewpoint, see the books by Kot and Murray [107, 150].

DEFINITION 1.1 *Let $I \subseteq \mathbb{R}$ be an interval (finite or infinite). We say a finite-valued function $u : I \to \mathbb{R}$ is piecewise continuous if it is continuous at each $t \in I$, with the possible exception of at most a finite number of t, and if u is equal to either its left or right limit at every $t \in I$.*

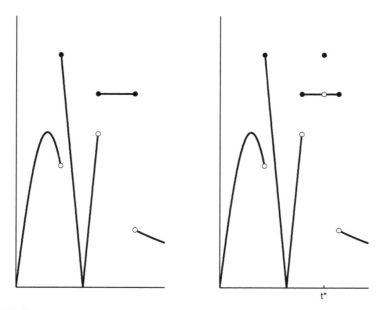

FIGURE 1.2: The graph to the left is an example of a piecewise continuous function. The graph to the right is not, because the value of the function at t^* is not the left or right limit.

Although somewhat nonstandard terminology, requiring piecewise continuous functions to equal their left or right limits eliminates a great many headaches farther down the road. In words, a piecewise continuous function can have finitely many "jump discontinuities" from one continuous segment to another. It cannot have a value that is an isolated single point (Figure 1.2).

Suppose $u : I \to \mathbb{R}$ is piecewise continuous. Let $g : \mathbb{R}^3 \to \mathbb{R}$ be continuous in three variables. Then, by the solution x of the differential equation

$$x'(t) = g(t, x(t), u(t)) \tag{1.1}$$

it is meant a continuous function $x : I \to \mathbb{R}$ which is differentiable, with x' satisfying the above expression, wherever u is continuous. Equivalently, if $I = [a, b]$, then x satisfies

$$x(t) = x(a) + \int_a^t g(s, x(s), u(s))\, ds.$$

An initial condition for $x(a)$ will normally be specified.

DEFINITION 1.2 *Let $x : I \to \mathbb{R}$ be continuous on I and differentiable at all but finitely points of I. Further, suppose that x' is continuous wherever it is defined. Then, we say x is piecewise differentiable.*

Note, if u is piecewise continuous, and x satisfies (1.1), then x is piecewise differentiable. Also, the actual value of u at its discontinuities is irrelevant in determining x. Throughout this text, all controls considered will be piecewise continuous, and we will not be concerned with values at discontinuities.

DEFINITION 1.3 *Let $k : I \to \mathbb{R}$. We say k is continuously differentiable if k' exists and is continuous on I.*

DEFINITION 1.4 *A function $k(t)$ is said to be concave on $[a, b]$ if*

$$\alpha k(t_1) + (1 - \alpha)k(t_2) \leq k\big(\alpha t_1 + (1 - \alpha)t_2\big)$$

for all $0 \leq \alpha \leq 1$ and for any $a \leq t_1, t_2 \leq b$

A function k is said to be *convex* on $[a, b]$ if it satisfies the reverse inequality, or equivalently, if $-k$ is concave. The second derivative of a twice differentiable concave function is non-positive; relating this to terminology used in calculus, concave here is "concave down" and convex is "concave up." If k is concave and differentiable, then we have a tangent line property

$$k(t_2) - k(t_1) \geq (t_2 - t_1)k'(t_2)$$

for all $a \leq t_1, t_2 \leq b$. In words, the slope of the secant line joining two points is less than the slope of the tangent line at the left point, and greater than the slope of the tangent line at the right point. See Figure 1.3.

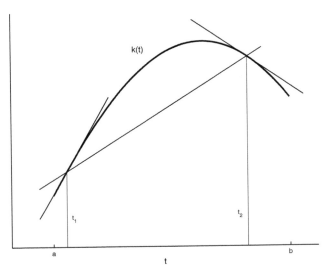

FIGURE 1.3: The graph of a concave function $k(t)$. The secant line and tangent lines for two points t_1 and t_2 are shown.

Analogously, a function $k(x, y)$ in two variables is said to be concave if

$$\alpha k(x_1, y_1) + (1 - \alpha) k(x_2, y_2) \leq k(\alpha x_1 + (1 - \alpha) x_2, \alpha y_1 + (1 - \alpha) y_2)$$

for all $0 \leq \alpha \leq 1$ and all (x_1, y_1), (x_2, y_2) in the domain of k. If k is such a function and has partial derivatives everywhere, then the analogue to the tangent line property is

$$k(x_1, y_1) - k(x_2, y_2) \geq (x_1 - x_2) k_x(x_1, y_1) + (y_1 - y_2) k_y(x_1, y_1)$$

for all pairs of points (x_1, y_1), (x_2, y_2) in the domain of k.

DEFINITION 1.5 *A function k is called Lipschitz if there exists a constant c (particular to k) such that $|k(t_1) - k(t_2)| \leq c|t_1 - t_2|$ for all points t_1, t_2 in the domain of k. The constant c is called the Lipschitz constant of k.*

THEOREM 1.1 (Mean Value Theorem)
Let k be continuous on $[a,b]$ and differentiable on (a,b). Then, there is some $x_0 \in (a,b)$ such that $k(b) - k(a) = k'(x_0)(b-a)$.

Note that a Lipschitz function is automatically continuous and, in fact, uniformly continuous. As such, this property is sometimes referred to as *Lipschitz continuity*. It follows from an application of the mean value theorem that if a function $k : I \to \mathbb{R}$ is piecewise differentiable on a bounded interval I, then k is Lipschitz.

1.2   The Basic Problem and Necessary Conditions

In our basic optimal control problem for ordinary differential equations, we use $u(t)$ for the control and $x(t)$ for the state. The state variable satisfies a differential equation which depends on the control variable:

$$x'(t) = g(t, x(t), u(t)).$$

As the control function is changed, the solution to the differential equation will change. Thus, we can view the control-to-state relationship as a map $u(t) \mapsto x = x(u)$ (of course, x is really a function of the independent variable t; we write $x(u)$ simply to remind us of the dependence on u). Our basic optimal control problem consists of finding a piecewise continuous control $u(t)$ and the associated state variable $x(t)$ to maximize the given objective functional, i.e.,

$$\max_u \int_{t_0}^{t_1} f(t, x(t), u(t)) \, dt$$

$$\text{subject to } \quad x'(t) = g(t, x(t), u(t))$$
$$x(t_0) = x_0 \text{ and } x(t_1) \text{ free.} \tag{1.2}$$

Such a maximizing control is called an optimal control. By $x(t_1)$ free, it is meant that the value of $x(t_1)$ is unrestricted. For our purposes, f and g will always be continuously differentiable functions in all three arguments. Thus, as the control(s) will always be piecewise continuous, the associated states will always be piecewise differentiable.

The principle technique for such an optimal control problem is to solve a set of "necessary conditions" that an optimal control and corresponding state must satisfy. It is important to understand the logical difference between necessary conditions and sufficient conditions of solution sets.

Necessary Conditions : If $u^*(t)$, $x^*(t)$ are optimal, then the following conditions hold ...

Sufficient Conditions : If $u^*(t)$, $x^*(t)$ satisfy the following conditions ..., then $u^*(t)$, $x^*(t)$ are optimal.

We will discuss sufficient conditions in the next chapter. For now, let us derive the necessary conditions. Express our objective functional in terms of the control:

$$J(u) = \int_{t_0}^{t_1} f(t, x(t), u(t)) \, dt,$$

where $x = x(u)$ is the corresponding state.

The necessary conditions that we derive were developed by Pontryagin and his co-workers in Moscow in the 1950's [158]. Pontryagin introduced the idea of "adjoint" functions to append the differential equation to the objective functional. Adjoint functions have a similar purpose as Lagrange multipliers in multivariate calculus, which append constraints to the function of several variables to be maximized or minimized. Thus, we begin by finding appropriate conditions that the adjoint function should satisfy. Then, by differentiating the map from the control to the objective functional, we will derive a characterization of the optimal control in terms of the optimal state and corresponding adjoint. So do not feel as if we are "pulling a rabbit out of the hat" when we define the adjoint equation.

FIGURE 1.4: Pulling the adjoint out of the hat.

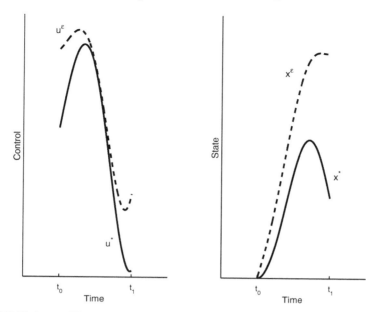

FIGURE 1.5: The optimal control u^* and state x^* (in solid) plotted together with u^ϵ and x^ϵ (dashed).

Assume a (piecewise continuous) optimal control exists, and that u^* is such a control, with x^* the corresponding state. Namely, $J(u) \leq J(u^*) < \infty$ for all controls u. Let $h(t)$ be a piecewise continuous variation function and $\epsilon \in \mathbb{R}$ a constant. Then

$$u^\epsilon(t) = u^*(t) + \epsilon h(t)$$

is another piecewise continuous control.

Let x^ϵ be the state corresponding to the control u^ϵ, namely, x^ϵ satisfies

$$\frac{d}{dt} x^\epsilon(t) = g(t, x^\epsilon(t), u^\epsilon(t)) \tag{1.3}$$

wherever u^ϵ is continuous. Since all trajectories start at the same position, we take $x^\epsilon(t_0) = x_0$ (Figure 1.5).

It is easily seen that $u^\epsilon(t) \to u^*(t)$ for all t as $\epsilon \to 0$. Further, for all t

$$\left.\frac{\partial u^\epsilon(t)}{\partial \epsilon}\right|_{\epsilon=0} = h(t).$$

In fact, something similar is true for x^ϵ. Because of the assumptions made on g, it follows that

$$x^\epsilon(t) \to x^*(t)$$

for each fixed t. Further, the derivative

$$\left.\frac{\partial}{\partial \epsilon} x^\epsilon(t)\right|_{\epsilon=0}$$

exists for each t. The actual value of quantity will prove unimportant. We need only to know that it exists.

The objective functional at u^ϵ is

$$J(u^\epsilon) = \int_{t_0}^{t_1} f(t, x^\epsilon(t), u^\epsilon(t))\, dt.$$

We are now ready to introduce the adjoint function or variable λ. Let $\lambda(t)$ be a piecewise differentiable function on $[t_0, t_1]$ to be determined. By the Fundamental Theorem of Calculus,

$$\int_{t_0}^{t_1} \frac{d}{dt}[\lambda(t) x^\epsilon(t)]\, dt = \lambda(t_1) x^\epsilon(t_1) - \lambda(t_0) x^\epsilon(t_0),$$

which implies

$$\int_{t_0}^{t_1} \frac{d}{dt}[\lambda(t) x^\epsilon(t)]\, dt + \lambda(t_0) x_0 - \lambda(t_1) x^\epsilon(t_1) = 0.$$

Adding this 0 expression to our $J(u^\epsilon)$ gives

$$\begin{aligned} J(u^\epsilon) &= \int_{t_0}^{t_1} \left[f(t, x^\epsilon(t), u^\epsilon(t)) + \frac{d}{dt}(\lambda(t) x^\epsilon(t)) \right] dt \\ &\quad + \lambda(t_0) x_0 - \lambda(t_1) x^\epsilon(t_1) \\ &= \int_{t_0}^{t_1} \left[f(t, x^\epsilon(t), u^\epsilon(t)) + \lambda'(t) x^\epsilon(t) + \lambda(t) g(t, x^\epsilon(t), u^\epsilon(t)) \right] dt \\ &\quad + \lambda(t_0) x_0 - \lambda(t_1) x^\epsilon(t_1), \end{aligned}$$

where we used the product rule and the fact that $g(t, x^\epsilon, u^\epsilon) = \frac{d}{dt} x^\epsilon$ at all but finitely many points. Since the maximum of J with respect to the control u occurs at u^*, the derivative of $J(u^\epsilon)$ with respect to ϵ (in the direction h) is zero, i.e.,

$$0 = \left.\frac{d}{d\epsilon} J(u^\epsilon)\right|_{\epsilon=0} = \lim_{\epsilon \to 0} \frac{J(u^\epsilon) - J(u^*)}{\epsilon}.$$

This gives a limit of an integral expression. A version of the Lebesgue Dominated Convergence Theorem [162, 163, 171] allows us to move the limit (and thus the derivative) inside the integral. This is due to the compact interval of integration and the piecewise differentiability of the integrand. Therefore,

$$0 = \frac{d}{d\epsilon} J(u^\epsilon) \Big|_{\epsilon=0}$$
$$= \int_{t_0}^{t_1} \frac{\partial}{\partial \epsilon} \Big[f(t, x^\epsilon(t), u^\epsilon(t)) + \lambda'(t) x^\epsilon(t) + \lambda(t) g(t, x^\epsilon(t), u^\epsilon(t)) \, dt \Big] \Big|_{\epsilon=0}$$
$$- \frac{\partial}{\partial \epsilon} \lambda(t_1) x^\epsilon(t_1) \Big|_{\epsilon=0}.$$

Applying the chain rule to f and g, it follows

$$0 = \int_{t_0}^{t_1} \left[f_x \frac{\partial x^\epsilon}{\partial \epsilon} + f_u \frac{\partial u^\epsilon}{\partial \epsilon} + \lambda'(t) \frac{\partial x^\epsilon}{\partial \epsilon} + \lambda(t) \left(g_x \frac{\partial x^\epsilon}{\partial \epsilon} + g_u \frac{\partial u^\epsilon}{\partial \epsilon} \right) \right] \Big|_{\epsilon=0} dt \quad (1.4)$$
$$- \lambda(t_1) \frac{\partial x^\epsilon}{\partial \epsilon}(t_1) \Big|_{\epsilon=0},$$

where the arguments of the f_x, f_u, g_x, and g_u terms are $(t, x^*(t), u^*(t))$. Rearranging the terms in (1.4) gives

$$0 = \int_{t_0}^{t_1} \left[\left(f_x + \lambda(t) g_x + \lambda'(t) \right) \frac{\partial x^\epsilon}{\partial \epsilon}(t) \Big|_{\epsilon=0} + (f_u + \lambda(t) g_u) h(t) \right] dt \quad (1.5)$$
$$- \lambda(t_1) \frac{\partial x^\epsilon}{\partial \epsilon}(t_1) \Big|_{\epsilon=0}.$$

We want to choose the adjoint function to simplify (1.5) by making the coefficients of

$$\frac{\partial x^\epsilon}{\partial \epsilon}(t) \Big|_{\epsilon=0}$$

vanish. Thus, we choose the adjoint function $\lambda(t)$ to satisfy

$$\lambda'(t) = -[f_x(t, x^*(t), u^*(t)) + \lambda(t) g_x(t, x^*(t), u^*(t))] \quad \text{(adjoint equation)},$$

and the boundary condition

$$\lambda(t_1) = 0 \quad \text{(transversality condition)}.$$

Now (1.5) reduces to

$$0 = \int_{t_0}^{t_1} \Big(f_u(t, x^*(t), u^*(t)) + \lambda(t) g_u(t, x^*(t), u^*(t)) \Big) h(t) \, dt.$$

As this holds for any piecewise continuous variation function $h(t)$, it holds for

$$h(t) = f_u(t, x^*(t), u^*(t)) + \lambda(t) g_u(t, x^*(t), u^*(t)).$$

In this case

$$0 = \int_{t_0}^{t_1} \Big(f_u(t, x^*(t), u^*(t)) + \lambda(t) g_u(t, x^*(t), u^*(t)) \Big)^2 dt,$$

which implies the *optimality condition*

$$f_u(t, x^*(t), u^*(t)) + \lambda(t) g_u(t, x^*(t), u^*(t)) = 0 \quad \text{for all } t_0 \leq t \leq t_1.$$

These equations form a set of necessary conditions that an optimal control and state must satisfy. In practice, one does not need to rederive the above equations in this way for a particular problem. In fact, we can generate the above necessary conditions from the Hamiltonian H, which is defined as follows,

$$H(t, x, u, \lambda) = f(t, x, u) + \lambda g(t, x, u)$$
$$= \text{integrand} + \text{adjoint} * \text{RHS of DE}.$$

We are maximizing H with respect to u at u^*, and the above conditions can be written in terms of the Hamiltonian:

$$\frac{\partial H}{\partial u} = 0 \text{ at } u^* \Rightarrow f_u + \lambda g_u = 0 \quad \textit{(optimality condition)},$$

$$\lambda' = -\frac{\partial H}{\partial x} \Rightarrow \lambda' = -(f_x + \lambda g_x) \quad \textit{(adjoint equation)},$$

$$\lambda(t_1) = 0 \quad \textit{(transversality condition)}.$$

We are given the dynamics of the state equation:

$$x' = g(t, x, u) = \frac{\partial H}{\partial \lambda}, \quad x(t_0) = x_0.$$

1.3 Pontryagin's Maximum Principle

These conclusions can be extended to a version of Pontryagin's Maximum Principle [158].

THEOREM 1.2

If $u^*(t)$ and $x^*(t)$ are optimal for problem (1.2), then there exists a piecewise differentiable adjoint variable $\lambda(t)$ such that

$$H(t, x^*(t), u(t), \lambda(t)) \leq H(t, x^*(t), u^*(t), \lambda(t))$$

for all controls u at each time t, where the Hamiltonian H is

$$H = f(t, x(t), u(t)) + \lambda(t) g(t, x(t), u(t)),$$

and

$$\lambda'(t) = -\frac{\partial H(t, x^*(t), u^*(t), \lambda(t))}{\partial x},$$
$$\lambda(t_1) = 0.$$

We have already shown with this adjoint and Hamiltonian, $H_u = 0$ at u^* for each t. Namely, the Hamiltonian has a critical point, in the u variable, at u^* for each t. It is not surprising that this critical point is a maximum considering the optimal control problem. However, the proof of this theorem is quite technical and difficult, and we omit it here. We refer the interested reader to Pontryagin's original text [158] and to Clarke's book for extensions [35]. The earlier requirement of controls being everywhere equal to either their left or right limits plays a pivotal role in the proof. Here, we state and prove the result for a very specific case, for illustrative purposes.

THEOREM 1.3

Suppose that $f(t, x, u)$ and $g(t, x, u)$ are both continuously differentiable functions in their three arguments and concave in u. Suppose u^* is an optimal control for problem (1.2), with associated state x^*, and λ a piecewise differentiable function with $\lambda(t) \geq 0$ for all t. Suppose for all $t_0 \leq t \leq t_1$

$$0 = H_u(t, x^*(t), u^*(t), \lambda(t)).$$

Then for all controls u and each $t_0 \leq t \leq t_1$, we have

$$H(t, x^*(t), u(t), \lambda(t)) \leq H(t, x^*(t), u^*(t), \lambda(t)).$$

PROOF Fix a control u and a point in time $t_0 \leq t \leq t_1$. Then,

$$H(t,x^*(t), u^*(t), \lambda(t)) - H(t, x^*(t), u(t), \lambda(t))$$
$$= \Big[f(t, x^*(t), u^*(t)) + \lambda(t) g(t, x^*(t), u^*(t))\Big]$$
$$\quad - \Big[f(t, x^*(t), u(t)) + \lambda(t) g(t, x^*(t), u(t))\Big]$$
$$= \Big[f(t, x^*(t), u^*(t)) - f(t, x^*(t), u(t))\Big]$$
$$\quad + \lambda(t) \Big[g(t, x^*(t), u^*(t)) - g(t, x^*(t), u(t))\Big]$$
$$\geq \big(u^*(t) - u(t)\big) f_u(t, x^*(t), u^*(t)) + \lambda(t)\big(u^*(t) - u(t)\big) g_u(t, x^*(t), u^*(t))$$
$$= \big(u^*(t) - u(t)\big) H_u(t, x^*(t), u^*(t), \lambda(t)) = 0.$$

The transition from line 3 to line 4 is attained from applying the tangent line property to f and g, and because $\lambda(t) \geq 0$. □

An identical argument generates the same necessary conditions when the problem is minimization rather than maximization. In a minimization problem, we are minimizing the Hamiltonian pointwise, and the inequality in Pontryagin's Maximum Principle in reversed. Indeed, for a minimization problem with f, g being convex in u, we can derive

$$H(t, x^*(t), u(t), \lambda(t)) \geq H(t, x^*(t), u^*(t), \lambda(t))$$

by the same argument as in Theorem 1.3.

We have converted the problem of finding a control that maximizes (or minimizes) the objective functional subject to the differential equation and initial condition, to maximizing the Hamiltonian pointwise with respect to the control. Thus to find the necessary conditions, we do not need to calculate the integral in the objective functional, but only use the Hamiltonian. Later, we will see the usefulness of the property that the Hamiltonian is maximized pointwise by an optimal control.

We can also check concavity conditions to distinguish between controls that maximize and those that minimize the objective functional [62]. If

$$\frac{\partial^2 H}{\partial u^2} < 0 \quad \text{at } u^*,$$

then the problem is maximization, while

$$\frac{\partial^2 H}{\partial u^2} > 0 \quad \text{at } u^*.$$

goes with minimization.

We can view our optimal control problem as having two unknowns, u^* and x^*, at the start. We have introduced an adjoint variable λ, which is similar to a Lagrange multiplier. It attaches the differential equation information onto the maximization of the objective functional. The following is an outline of how this theory can be applied to solve the simplest problems.

1. Form the Hamiltonian for the problem.

2. Write the adjoint differential equation, transversality boundary condition, and the optimality condition. Now there are three unknowns, u^*, x^*, and λ.

3. Try to eliminate u^* by using the optimality equation $H_u = 0$, i.e., solve for u^* in terms of x^* and λ.

4. Solve the two differential equations for x^* and λ with two boundary conditions, substituting u^* in the differential equations with the expression for the optimal control from the previous step.

5. After finding the optimal state and adjoint, solve for the optimal control.

If the Hamiltonian is linear in the control variable u, it can be difficult to solve for u^* from the optimality equation; we will treat this case in Chapter 17. If we can solve for u^* from the optimality equation, we are then left with two unknowns x^* and λ satisfying two differential equations with two boundary conditions. We solve that system of differential equations for the optimal state and adjoint and then obtain the optimal control. We will see in some simple examples that the system can be solved analytically (by hand) and in other examples that the system can be solved numerically.

When we are able to solve for the optimal control in terms of x^* and λ, we will call that formula for u^* the *characterization of the optimal control*. The state equations and the adjoint equations together with the characterization of the optimal control and the boundary conditions are called the *optimality system*. For now, let us try to better understand these ideas with a few examples.

Example 1.1   (from [100])

$$\min_u \int_0^1 u(t)^2 \, dt$$

subject to $x'(t) = x(t) + u(t)$, $x(0) = 1$, $x(1)$ free.

Can we see what the optimal control should be? The goal of the problem is to minimize this integral, which does not involve the state. Only the integral of control (squared) is to be minimized. Therefore, we expect the optimal control is 0. We verify with the necessary conditions.

We begin by forming the Hamiltonian H

$$H = u^2 + \lambda(x + u).$$

The optimality condition is

$$0 = \frac{\partial H}{\partial u} = 2u + \lambda \text{ at } u^* \Rightarrow u^* = -\frac{1}{2}\lambda.$$

We see the problem is indeed minimization as

$$\frac{\partial^2 H}{\partial u^2} = 2 > 0.$$

The adjoint equation is given by

$$\lambda' = -\frac{\partial H}{\partial x} = -\lambda \Rightarrow \lambda(t) = ce^{-t},$$

for some constant c. But, the transversality condition is

$$\lambda(1) = 0 \Rightarrow ce^{-1} = 0 \Rightarrow c = 0.$$

Thus, $\lambda \equiv 0$, so that $u^* = -\lambda/2 = 0$. So, x^* satisfies $x' = x$ and $x(0) = 1$. Hence, the optimal solutions are

$$\lambda \equiv 0, \quad u^* \equiv 0, \quad x^*(t) = e^t,$$

and the state function is plotted in Figure 1.6.

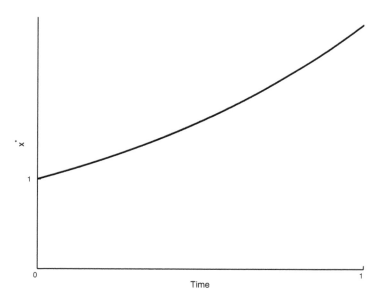

FIGURE 1.6: Optimal state for Example 1.1 plotted as a function of time.

Example 1.2

$$\min_u \frac{1}{2} \int_0^1 3x(t)^2 + u(t)^2 \, dt$$

subject to $\quad x'(t) = x(t) + u(t), \; x(0) = 1.$

The $\frac{1}{2}$ which appears before the integral will have no effect on the minimizing control and, thus, no effect on the problem. It is inserted in order to make the computations slightly neater. You will see how shortly. Also, note we have omitted the phrase "$x(1)$ free" from the statement of the problem. This is standard notation, in that a term which is unrestricted is simply not mentioned. We adopt this convention from now on.

Form the Hamiltonian of the problem

$$H = \frac{3}{2}x^2 + \frac{1}{2}u^2 + x\lambda + u\lambda.$$

The optimality condition gives

$$0 = \frac{\partial H}{\partial u} = u + \lambda \text{ at } u^* \Rightarrow u^* = -\lambda.$$

Notice $\frac{1}{2}$ cancels with the 2 which comes from the square on the control u. Also, the problem is a minimization problem as

$$\frac{\partial^2 H}{\partial u^2} = 1 > 0.$$

We use the Hamiltonian to find a differential equation of the adjoint λ,

$$\lambda'(t) = -\frac{\partial H}{\partial x} = -3x - \lambda, \; \lambda(1) = 0.$$

Substituting the derived characterization for the control variable u in the equation for x', we arrive at

$$\begin{pmatrix} x \\ \lambda \end{pmatrix}' = \begin{pmatrix} 1 & -1 \\ -3 & -1 \end{pmatrix} \begin{pmatrix} x \\ \lambda \end{pmatrix}.$$

The eigenvalues of the coefficient matrix are 2 and -2. Finding the eigenvectors, the equations for x and λ are

$$\begin{pmatrix} x \\ \lambda \end{pmatrix}(t) = c_1 \begin{pmatrix} 1 \\ -1 \end{pmatrix} e^{2t} + c_2 \begin{pmatrix} 1 \\ 3 \end{pmatrix} e^{-2t}.$$

Using $x(0) = 1$ and $\lambda(1) = 0$, we find $c_1 = 3c_2 e^{-4}$ and $c_2 = \frac{1}{3e^{-4}+1}$. Thus, using the optimality equation, the optimal solutions are

$$u^*(t) = \frac{3e^{-4}}{3e^{-4}+1}e^{2t} - \frac{3}{3e^{-4}+1}e^{-2t},$$

$$x^*(t) = \frac{3e^{-4}}{3e^{-4}+1}e^{2t} + \frac{1}{3e^{-4}+1}e^{-2t},$$

which are illustrated in Figure 1.7.

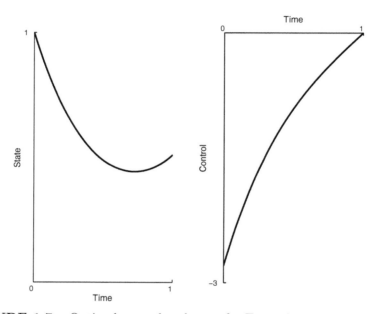

FIGURE 1.7: Optimal control and state for Example 1.2.

1.4 Exercises

In the following exercises, write out the necessary conditions for each problem, then solve the optimality system (unless otherwise stated) to find the optimal control and state.

Exercise 1.1 Write out the necessary conditions for the following problem to be treated in Lab 1. Do not attempt to solve the resulting optimality system.

$$\max_u \int_0^1 Ax(t) - Bu(t)^2 \, dt$$

subject to $x'(t) = -\frac{1}{2}x(t)^2 + Cu(t)$, $x(0) = x_0 > -2$,

$A \geq 0, B > 0.$

Exercise 1.2 Solve

$$\min_u \int_1^2 tu(t)^2 + t^2 x(t) \, dt$$

subject to $x'(t) = -u(t)$, $x(1) = 1$.

Exercise 1.3 (from [100]) Solve

$$\max_u \int_1^5 u(t)x(t) - u(t)^2 - x(t)^2 \, dt$$

subject to $x'(t) = x(t) + u(t)$, $x(1) = 2$.

Exercise 1.4 (from [100]) Solve

$$\min_u \int_0^1 x(t)^2 + x(t) + u(t)^2 + u(t) \, dt$$

subject to $x'(t) = u(t)$, $x(0) = 0$.

Exercise 1.5 Let $y(t) = t + 1$. Solve

$$\min_u \frac{1}{2}\int_0^1 (x(t) - y(t))^2 + u(t)^2 \, dt$$

subject to $x'(t) = u(t)$, $x(0) = 1$.

Exercise 1.6 Formulate an optimal control problem for a population with an Allee effect growth term, in which the control is the proportion of the population to be harvested. This means that differential equation has an Allee effect term. Choose an objective functional which maximizes revenue from the harvesting while minimizing the cost of harvesting. The revenue is the integral of amount harvested per time. Assume the cost of harvesting has a quadratic format. See [107] for the terminology like "Allee."

Chapter 2

Existence and Other Solution Properties

In the last chapter, we developed necessary conditions to solve basic optimal control problems. However, some difficulties can arise with this method. It is possible that the necessary conditions could yield multiple solution sets, only some of which are optimal controls. Further, recall that in the development of the necessary conditions, we began by assuming an optimal control exists. It is also possible that the necessary conditions could be solvable when the original optimal control problem has no solution. We expect the objective functional evaluated at the optimal state and control to give a finite answer. If this objective functional value turns out to be ∞ or $-\infty$, we would say the problem has no solution. An example of this is given below.

Example 2.1 (from [100])

$$\max_u \int_0^1 x(t) + u(t)\, dt$$

subject to $x'(t) = 1 - u(t)^2,\ x(0) = 1.$

The Hamiltonian and the optimality condition are:

$$H(t, x, u, \lambda) = x + u + \lambda(1 - u^2),$$
$$\frac{\partial H}{\partial u} = 1 - 2\lambda u = 0 \Rightarrow u = \frac{1}{2\lambda}.$$

From the adjoint equation and its boundary condition,

$$\lambda' = -\frac{\partial H}{\partial x} = -1 \text{ and } \lambda(1) = 0,$$

we can directly calculate

$$\lambda(t) = 1 - t.$$

Note that the concavity with respect to the control u is correct for a maximization problem,

$$H_{uu} = -2\lambda \leq 0,$$

as $\lambda(t) \geq 0$. Next, we calculate the optimal state using the differential equation and its boundary condition

$$x' = 1 - u^2 = 1 - \frac{1}{4(1-t)^2} \quad \text{and} \quad x(0) = 1,$$

and find that

$$x^*(t) = t - \frac{1}{4(1-t)} + \frac{5}{4} \quad \text{and} \quad u^*(t) = \frac{1}{2(1-t)}.$$

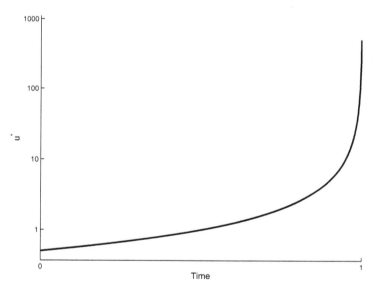

FIGURE 2.1: The graph of u^* from Example 2.1, plotted in logarithmic scale. The value of u^* tends to infinity as t approaches 1.

However, notice that when the objective functional is evaluated at the optimal control and state, we do not obtain a finite answer,

$$\int_0^1 [x^*(t) + u^*(t)]\, dt = \int_0^1 t + \frac{5}{4} + \frac{1}{4(1-t)}\, dt = \infty.$$

There is not an "optimal control" in this case, since we are considering problems with finite maximum (or minimum) objective functional values, even though the solutions we found satisfy the necessary conditions. In this simple example, the optimal control and state become unbounded at $t = 1$. See Figure 2.1. In most applications, we want the optimal control and state values to remain bounded. Later we will consider the restriction of imposing a bound on the control set in the setting of the problem.

What causes this unbounded difficulty in this example? One explanation is the quadratic nonlinearity u^2 in the differential equation. For example, consider the simple differential equation

$$x' = x^2 \quad \text{with} \quad x(0) = 1,$$

which has a quadratic term. The solution is easily shown to be

$$x = \frac{1}{1-t},$$

which become unbounded in finite time (at $t = 1$). This illustrates how difficulty can arise with quadratic terms in differential equations.

2.1 Existence and Uniqueness Results

It should be clear now that simply solving the necessary conditions is not enough to solve an optimal control problem. To completely justify the methods used in Chapter 1, some existence and uniqueness results should be examined. First, what conditions can guarantee the existence of a finite objective functional value at the optimal control and state? We state some simple results for existence from Fleming and Rishel, Kamien and Schwartz, and Macki and Strauss [62, 100, 135]. For more cases, see Cesari [29]. The following is an example of a sufficient condition result.

THEOREM 2.1
Consider

$$J(u) = \int_{t_0}^{t_1} f(t, x(t), u(t))\, dt$$

subject to $\quad x'(t) = g(t, x(t), u(t)), \ x(t_0) = x_0.$

Suppose that $f(t, x, u)$ and $g(t, x, u)$ are both continuously differentiable functions in their three arguments and concave in x and u. Suppose u^* is a control, with associated state x^*, and λ a piecewise differentiable function, such that u^*, x^*, and λ together satisfy on $t_0 \le t \le t_1$:

$$fu + \lambda g_u = 0,$$
$$\lambda' = -(f_x + \lambda g_x),$$
$$\lambda(t_1) = 0,$$
$$\lambda(t) \geq 0.$$

Then for all controls u, we have

$$J(u^*) \geq J(u).$$

PROOF Let u be any control, and x its associated state. Note, as $f(t, x, u)$ is concave in both the x and u variable, we have by the tangent line property (see Section 1.1)

$$f(t, x^*, u^*) - f(t, x, u) \geq (x^* - x)f_x(t, x^*, u^*) + (u^* - u)f_u(t, x^*, u^*).$$

This gives

$$\begin{aligned} J(u^*) - J(u) &= \int_{t_0}^{t_1} f(t, x^*, u^*) - f(t, x, u)\, dt \\ &\geq \int_{t_0}^{t_1} (x^*(t) - x(t))f_x(t, x^*, u^*) \\ &\quad + (u^*(t) - u(t))f_u(t, x^*, u^*)\, dt. \end{aligned} \qquad (2.1)$$

Substituting

$$f_x(t, x^*, u^*) = -\lambda'(t) - \lambda(t)g_x(t, x^*, u^*) \quad \text{and}$$
$$f_u(t, x^*, u^*) = -\lambda(t)g_u(t, x^*, u^*),$$

as given by the hypothesis, the last term in (2.1) becomes

$$\int_{t_0}^{t_1} (x^*(t) - x(t))(-\lambda(t)g_x(t, x^*, u^*) - \lambda'(t)) \\ + (u^*(t) - u(t))(-\lambda(t)g_u(t, x^*, u^*))\, dt.$$

Using integration by parts, and recalling $\lambda(t_1) = 0$ and $x(t_0) = x^*(t_0)$, we see

$$\int_{t_0}^{t_1} -\lambda'(t)(x^*(t) - x(t))\, dt = \int_{t_0}^{t_1} \lambda(t)(x^*(t) - x(t))'\, dt = \\ \int_{t_0}^{t_1} \lambda(t)\big(g(t, x^*(t), u^*(t)) - g(t, x(t), u(t))\big)\, dt.$$

Making this substitution,

$$J(u^*) - J(u) \geq \int_{t_0}^{t_1} \lambda(t)\Big[g(t, x^*, u^*) - g(t, x, u) -$$
$$(x^* - x)g_x(t, x^*, u^*) - (u^* - u)g_u(t, x^*, u^*)\Big] dt.$$

Taking into account $\lambda(t) \geq 0$ and that g is concave in both x and u, this gives the desired result $J(u^*) - J(u) \geq 0$. □

Note that the Example 2.1 satisfies the conditions for this theorem and this conclusion, but $J(u^*)$ is not a finite. A true existence result guarantees an optimal control, with finite objective functional. Such results usually require some restrictions on f and/or g. Here is an example of an existence result from [62].

THEOREM 2.2
Let the set of controls for problem (1.2) be Lebesgue integrable functions (instead of just piecewise continuous functions) on $t_0 \leq t \leq t_1$ with values in $\mathbb{R}$. Suppose that $f(t, x, u)$ is convex in u, and there exist constants C_4 and $C_1, C_2, C_3 > 0$ and $\beta > 1$ such that

$$g(t, x, u) = \alpha(t, x) + \beta(t, x)u$$
$$|g(t, x, u)| \leq C_1(1 + |x| + |u|)$$
$$|g(t, x_1, u) - g(t, x, u)| \leq C_2|x_1 - x|(1 + |u|)$$
$$f(t, x, u) \geq C_3|u|^\beta - C_4$$

for all t with $t_0 \leq t \leq t_1$, x, x_1, u in $\mathbb{R}$. Then there exists an optimal control u^* maximizing $J(u)$, with $J(u^*)$ finite.

For a minimization problem, g would have a concave property and the inequality on f would be reversed. Note that Example 2.1 does not satisfy the first inequality assumption on g, nor the assumption on f.

You may be unfamiliar with the term Lebesgue integrable in the above theorem. This is a concept used in higher levels of analysis. It is sufficient here for you to know that all Riemann integrable functions are also Lebesgue integrable. There are many excellent sources pertaining to Lebesgue integration and measure theory. Royden [162], Rudin [163], and Stein and Shakarchi [171] are some of the widely-used textbooks. Further, note that the necessary conditions developed to this point deal with piecewise continuous optimal controls, while this existence theorem guarantees an optimal control which is only Lebesgue integrable. This disconnect can be remedied by extending the necessary conditions, in a meaningful way, to Lebesgue integrable functions [135], but we do not treat this idea here. See the existence of optimal control results in [58]. Convexity (or concavity, depending on the type of problem) in the control set is frequently used, but there are some results without convexity [16, 29, 58]. In addition, there are existence results for solutions of the state equation [131].

Also of interest is the idea of uniqueness. Suppose an optimal control exists, i.e, there is u^* such that $J(u) \leq J(u^*) < \infty$ for all controls u (in the maximization case). We say u^* is unique if whenever $J(u^*) = J(u)$, then $u^* = u$ at all but finitely many points. In this case, the associated states will be identical. We call this state, x^*, the unique optimal state.

Clearly, uniqueness of solutions of the optimality system implies uniqueness of the optimal control, if one exists. We can frequently prove uniqueness of the solutions of the optimality system, but only for a small time interval. This small time condition is due to opposite time orientations of the state equation and adjoint equation, meaning the state equation has an initial time condition and the adjoint equation has a final time condition.

However, in general, uniqueness of the optimal control does not necessarily guarantee uniqueness of the optimality system. To prove uniqueness of the optimal control directly, strict concavity of the objective functional $J(u, x(u))$ must be established. Direct uniqueness results tend to be cumbersome and difficult to state, and, as they will not be needed here, they will not be treated.

If f, g, and the right hand side of the adjoint equation are Lipschitz in the state and adjoint variables, then the uniqueness of solutions of the optimality system holds for small time. Sometimes, if the solutions of the optimality system are bounded, then one can easily get the Lipschitz property and the resulting uniqueness [60].

We have chosen lab problems and examples (with the exception of Example 2.1) for this book that satisfy an existence result of some kind and have optimality systems which guarantee uniqueness of solutions for small enough time intervals. Therefore, it will be sufficient to solve only the necessary conditions, as has been done in the previous examples and will continue to be done.

2.2 Interpretation of the Adjoint

For an interpretation of the adjoint variable, we first define the value function V by

$$V(x_0, t_0) := \max_u \int_{t_0}^{t_1} f(t, x(t), u(t)) \, dt$$

$$\text{subject to} \quad x'(t) = g(t, x, u), \quad x(t_0) = x_0.$$

The value function gives the value of the integral when evaluated at the optimal control and state. The notation $V(x_0, t_0)$ indicates the dependence on the initial state and time. There is a relationship between the adjoint variable and the derivative of the value function with respect to the state, which can be stated as

$$\frac{\partial V}{\partial x}(x_0, t_0) = \lim_{\epsilon \to 0} \frac{V(x_0 + \epsilon, t_0) - V(x_0, t_0)}{\epsilon} = \lambda(t_0).$$

In the case that the objective functional represents profit or cost, then the units of $\frac{\partial V}{\partial x}$ are money per unit item. Thus, the adjoint variable $\lambda(t_0)$ is equal to the marginal variation in the value function with respect to the state at time t_0, and is commonly called the *shadow price*. One can view λ as the additional money (profit/cost) associated with an additional increment of the state variable. In fact, this interpretation is valid for all time t [100]:

$$\frac{\partial V}{\partial x}(x^*(t), t) = \lambda(t) \quad \text{for all} \quad t_0 \leq t \leq t_1.$$

If a fishery harvest problem was being considered, the state would represent the amount of fish and the objective functional represents the profit made at the fishery. In this case, one would view the adjoint variables as giving: "If one fish is added to the stock at time t_0, how much is the value of the fishery affected?" Using

$$\frac{\partial V}{\partial x}(x_0, t_0) = \lambda(t_0),$$

we can approximate the difference by

$$\frac{V(x_0 + \epsilon, t_0) - V(x_0, t_0)}{\epsilon} \approx \lambda(t_0) \Rightarrow$$
$$V(x_0 + \epsilon, t_0) \approx V(x_0, t_0) + \epsilon \lambda(t_0),$$

for all small $\epsilon > 0$. Just for the idea, take $\epsilon = 1$ and obtain

$$V(x_0 + 1, t_0) \approx V(x_0, t_0) + \lambda(t_0),$$

which means that the additional profit resulting from adding one more fish to the stock at the initial time is given by $\lambda(t_0)$. This interpretation can help one to know what sign (positive or negative) to expect from an adjoint variable in particular applications.

2.3 Principle of Optimality

An important result in both optimal control and dynamic programming is the Principle of Optimality. It concerns optimizing a system over a subinterval of the original time span, and in particular, how the optimal control over this smaller interval relates to the optimal control on the full time period.

THEOREM 2.3
Let u^* be an optimal control, and x^* the resulting state, for the problem

$$\max_u J(u) = \max_u \int_{t_0}^{t_1} f(t, x(t), u(t))\, dt$$

$$\text{subject to} \quad x'(t) = g(t, x(t), u(t)), \quad x(t_0) = x_0. \tag{2.2}$$

Let $\hat{t}$ be a fixed point in time such that $t_0 < \hat{t} < t_1$. Then, the restricted functions $\hat{u}^* = u^*|_{[\hat{t}, t_1]}$, $\hat{x}^* = x^*|_{[\hat{t}, t_1]}$, form an optimal pair for the restricted problem

$$\max_u \hat{J}(u) = \max_u \int_{\hat{t}}^{t_1} f(t, x(t), u(t))\, dt$$

$$\text{subject to} \quad x'(t) = g(t, x(t), u(t)), \quad x(\hat{t}) = x^*(\hat{t}). \tag{2.3}$$

Further, if u^* is the unique optimal control for (2.2), then $\hat{u}^*$ is the unique optimal control for (2.3).

PROOF This proof is done by contradiction. Suppose, to the contrary, that $\hat{u}^*$ is not optimal, i.e., there exists a control $\hat{u}_1$ on the interval $[\hat{t}, t_1]$ such that $\hat{J}(\hat{u}_1) > \hat{J}(\hat{u}^*)$. Construct a new control u_1 on the whole interval $[t_0, t_1]$ as follows

$$u_1(t) = \begin{cases} u^*(t) & \text{for } t_0 \leq t \leq \hat{t}, \\ \hat{u}_1(t) & \text{for } \hat{t} < t \leq t_1. \end{cases}$$

Let x_1 be the state associated with control u_1. Notice that u_1 and u^* agree on $[t_0, \hat{t}]$, so that x_1 and x^* will also agree there. Hence,

$$J(u_1) - J(u^*) = \left(\int_{t_0}^{\hat{t}} f(t, x_1, u_1)\, dt + \hat{J}(\hat{u}_1) \right) - \left(\int_{t_0}^{\hat{t}} f(t, x^*, u^*)\, dt + \hat{J}(\hat{u}^*) \right)$$

$$= \hat{J}(\hat{u}_1) - \hat{J}(\hat{u}^*)$$

$$> 0.$$

However, this contradicts our initial assumption that u^* was optimal for (2.2). Thus, no such control $\hat{u}_1$ exists, and $\hat{u}^*$ is optimal for (2.3).

The proof of the result concerning uniqueness follows in almost exactly the same manner and is left as an exercise. □

Note that this theorem also holds for minimization problems. Intuitively, the theorem makes sense. If we have an optimal pair u^*, x^* for an optimal control problem and move along the optimal path to a time $\hat{t}$, then one optimal path for the remaining time should be to simply continue on the path already begun. Notice, however, that when considering a time interval shortened by truncating the end, i.e., $[t_0, \hat{t}]$, we have no information on the new optimal control. In fact, no relation between the controls is necessary, as you will see in Example 2.3. First, let us study an example where the principle does apply.

Example 2.2

$$\min_u \int_0^2 x(t) + \frac{1}{2} u(t)^2 \, dt$$

subject to $\quad x'(t) = x(t) + u(t), \; x(0) = \frac{1}{2} e^2 - 1.$

First, we will solve this example on $[0, 2]$, then solve the same problem on a smaller interval $[1, 2]$. The Hamiltonian in this example is

$$H = x + \frac{1}{2} u^2 + x\lambda + u\lambda.$$

The adjoint equation and transversality condition give

$$\lambda' = -\frac{\partial H}{\partial x} = -1 - \lambda, \; \lambda(2) = 0 \;\Rightarrow\; \lambda(t) = e^{2-t} - 1,$$

and the optimality condition leads to

$$0 = \frac{\partial H}{\partial u} = u + \lambda \;\Rightarrow\; u^*(t) = -\lambda(t) = 1 - e^{2-t}.$$

Finally, from the state equation, the associated state is

$$x^*(t) = \frac{1}{2} e^{2-t} - 1.$$

Now, consider the same problem, except on the interval $[1, 2]$, i.e.,

$$\min_u \int_1^2 x(t) + \frac{1}{2} u(t)^2 \, dt$$

subject to $\quad x'(t) = x(t) + u(t), \; x(1) = \frac{1}{2} e - 1$

Clearly, the Principle of Optimality can be applied to find an optimal pair immediately, namely, the pair found above. The original problem on the interval $[0, 2]$ has the same optimal control as the above problem on $[1, 2]$. Let us solve this example by hand, though, to reinforce the power of the theorem. The Hamiltonian will be the same, regardless of interval. Because the end point remains fixed, the adjoint equation and transversality also remain the same:

$$\lambda' = -\frac{\partial H}{\partial x} = -1 - \lambda, \ \lambda(2) = 0 \ \Rightarrow \ \lambda(t) = e^{2-t} - 1,$$

while the optimality is also unchanged,

$$0 = \frac{\partial H}{\partial u} = u + \lambda \ \Rightarrow \ u^*(t) = -\lambda(t) = 1 - e^{2-t}.$$

Using the new initial condition $x(1) = \frac{1}{2}e - 1$, we find the corresponding state

$$x^*(t) = \frac{1}{2}e^{2-t} - 1.$$

Of course, we see the same optimal pair as above, as called for by the Principle of Optimality.

Now, let us examine another variation of the same problem. This time, we will consider the beginning of the interval $[0, 1]$. Notice that the original initial condition is used.

Example 2.3

$$\min_u \int_0^1 x(t) + \frac{1}{2}u(t)^2 \, dt$$

subject to $\quad x'(t) = x(t) + u(t), \ x(0) = \frac{1}{2}e^2 - 1.$

Again, the Hamiltonian is the same, so that the adjoint and optimality conditions are unchanged. However, the transversality condition is now different,

$$\lambda'(t) = -\frac{\partial H}{\partial x} = -1 - \lambda, \ \lambda(1) = 0 \ \Rightarrow \ \lambda(t) = e^{1-t} - 1,$$

so that

$$u^*(t) = -\lambda(t) = 1 - e^{1-t}.$$

Using this in the state equation,

$$x^*(t) = \frac{1}{2}e^{1-t} - 1 + \frac{1}{2}(e^2 - e)e^t.$$

Existence and Other Solution Properties 31

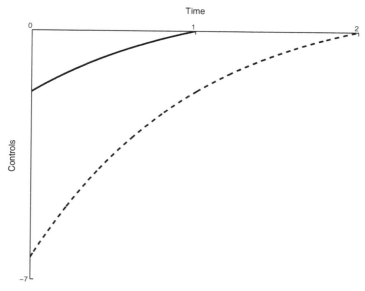

FIGURE 2.2: Optimal controls for Examples 2.2 (dashed) and 2.3 (solid) plotted together.

We notice that the optimal pair here is different from the optimal pair found in the previous example (Figure 2.2). This illustrates the limitations of Theorem 2.3. These examples also give an insight into the theorem via Pontryagin's Maximum Principle. When the initial time is increased, neither the adjoint equation nor the transversality condition is altered, so that the adjoint will remain the same. However, when the final time is decreased, the transversality condition will change. Then, the adjoint could be different, possibly changing the optimal pair u^*, x^*.

2.4 The Hamiltonian and Autonomous Problems

Recall that the Hamiltonian $H(t, x, u, \lambda)$ is a function of four variables. However, time t is the underlying variable as each of x, u, and λ is a function of t. Therefore, H can be thought of implicitly as a function of t. Because x and λ are continuous, and u is piecewise continuous, it follows H is a piecewise continuous function of t. In fact, a much stronger property is true for u^* and x^*.

THEOREM 2.4
The Hamiltonian is a Lipschitz continuous function of time t on the optimal path.

PROOF Let u^*, x^* be an optimal pair for (1.2), and λ the associated adjoint. For $t \in [t_0, t_1]$, write

$$M(t) = H(t, x^*(t), u^*(t), \lambda(t)).$$

As u^* is piecewise continuous on a compact interval, there is some bounded interval P such that $u^*(t) \in P$ for all $t \in [t_0, t_1]$. Similarly, there exist bounded intervals Q and R such that $x^*(t) \in Q$ and $\lambda(t) \in R$ for all $t \in [t_0, t_1]$.

Consider the Hamiltonian as a function of four variables $H(t, x, u, \lambda)$, where we think of x, u, λ as only numbers for a moment. By the original choices of f and g, H is continuously differentiable in all four arguments. Therefore, it is possible to choose a constant K_1 such that

$$|H_t(t, x, u, \lambda)| \leq K_1, \ |H_x(t, x, u, \lambda)| \leq K_1, \ \text{and} \ |H_\lambda(t, x, u, \lambda)| \leq K_1,$$

for all tuples (t, x, u, λ) in the compact set $[t_0, t_1] \times P \times Q \times R$. Fix $s, t \in [t_0, t_1]$. For convenience, write $x_t = x^*(t)$ and $x_s = x^*(s)$. Define $u_t, u_s, \lambda_t, \lambda_s$ similarly. Let $\tau \in P$. By a few applications of the mean value theorem,

$$\begin{aligned}|H(t, x_t, \tau, \lambda_t) &- H(s, x_s, \tau, \lambda_t)| \\ &\leq |H_t(c_1, x_t, \tau, \lambda_t)||t - s| + |H_x(s, c_2, \tau, \lambda_t)||x_t - x_s| \\ &\quad + |H_\lambda(s, x_s, \tau, c_3)||\lambda_t - \lambda_s| \\ &\leq K_1|t - s| + K_1|x_t - x_s| + K_1|\lambda_t - \lambda_s|,\end{aligned}$$

for some intermediary points $c_1 \in [t_0, t_1]$, $c_2 \in Q$, and $c_3 \in R$. On the other hand, x^* and λ are piecewise differentiable on a compact interval, thus Lipschitz continuous. Let K_2 be the maximum of the two Lipschitz constants. Then we have

$$\begin{aligned}|H(t, x_t, \tau, \lambda_t) &- H(s, x_s, \tau, \lambda_t)| \\ &\leq K_1|t - s| + K_1|x_t - x_s| + K_1|\lambda_t - \lambda_s| \\ &\leq (K_1 + 2K_1K_2)|t - s|.\end{aligned} \quad (2.4)$$

Set $K = K_1 + 2K_1K_2$ and note this holds for all $\tau \in P$.

Now, $M(t) = H(t, x_t, u_t, \lambda_t)$ and similarly for s. By Theorem 1.2, the Hamiltonian is maximized pointwise by u^*, so

$$\begin{aligned}H(t, x_t, u_s, \lambda_t) &\leq H(t, x_t, u_t, \lambda_t) \ \text{and} \\ H(s, x_s, u_t, \lambda_s) &\leq H(s, x_s, u_s, \lambda_s).\end{aligned} \quad (2.5)$$

Applying (2.4) for $\tau = u_s$ and $\tau = u_t$, and combining with (2.5), we see

$$\begin{aligned}-K|t-s| &\leq H(t, x_t, u_s, \lambda_t) - H(s, x_s, u_s, \lambda_s)\\ &\leq H(t, x_t, u_t, \lambda_t) - H(s, x_s, u_s, \lambda_s)\\ &= M(t) - M(s)\\ &\leq H(t, x_t, u_t, \lambda_t) - H(s, x_s, u_t, \lambda_s)\\ &\leq K|t-s|.\end{aligned}$$

Namely, $|M(t) - M(s)| \leq K|t-s|$. As t, s are arbitrary, M is Lipschitz continuous. □

In Exercise 1.2, we had a problem where f and g explicitly depend on t:

$$\min_u \int_1^2 tu(t)^2 + t^2 x(t)\, dt$$

$$\text{subject to}\quad x'(t) = -u(t),\ x(1) = 1.$$

The optimal solution set is

$$u^*(t) = \frac{1}{2t}\left(\frac{8}{3} - \frac{t^3}{3}\right),$$

$$x^*(t) = \frac{1}{18}t^3 - \frac{4}{3}\ln(t) + \frac{17}{18},$$

$$\lambda(t) = \frac{8}{3} - \frac{1}{3}t^3.$$

So, we can write

$$\begin{aligned}H(t, x^*, u^*, \lambda) &= tu^*(t)^2 + t^2 x^*(t) - \lambda(t)u^*(t)\\ &= -\frac{1}{4t}\left(\frac{8}{3} - \frac{1}{3}t^3\right)^2 + t^2\left(\frac{1}{18}t^3 - \frac{4}{3}\ln(t) + \frac{17}{18}\right),\end{aligned}$$

giving us H as an explicit function of t. Similarly, for Example 1.2, we can plug in the solved optimal solutions to find H as an explicit function of t. Amazingly, in this case

$$H(t) \equiv \frac{24e^{-4}}{(3e^{-4} + 1)^2},$$

a constant. This is not a coincidence. Before exploring this, we make a definition.

If an optimal control problem has no explicit dependence on time t, we say it is *autonomous*. In our notation, this means the function f (the integrand)

and g (the RHS of the state equation) are both functions of only x and u. Namely,

$$\max_u \int_{t_0}^{t_1} f(x(t), u(t))\, dt$$

subject to $\quad x'(t) = g(x(t), u(t)),\ x(0) = x_0.$ \hfill (2.6)

Examples 1.1 and 1.2 are both autonomous. Exercise 1.2 is not autonomous, as the integrand $f(t, x, u) = tu(t)^2 + t^2 x(t)$ has explicit dependence on t.

THEOREM 2.5
If an optimal control problem is autonomous, then the Hamiltonian is a constant function of time along the optimal path.

PROOF Let u^*, x^* be the optimal pair for the control problem (2.6), and λ the associated adjoint.

Let $M(t) = H(x^*(t), u^*(t), \lambda(t))$ be defined as in the proof of Theorem 2.4, except now H is only a function of three variables. As M is Lipschitz continuous, we have from measure theory that M is differentiable almost everywhere, with respect to Lebesgue measure [163]. Let $\bar{t} \in (t_0, t_1)$ be any point where M' exists.

Denote $u^*(\bar{t}) = \tau$. Note, for small enough $\delta > 0$ so that $\bar{t} + \delta \in [t_0, t_1]$, the Maximum Principle gives $M(\bar{t} + \delta) \geq H(x^*(\bar{t} + \delta), u^*(\bar{t}), \lambda(\bar{t} + \delta)) = H(x^*(\bar{t} + \delta), \tau, \lambda(\bar{t} + \delta))$. So,

$$M(\bar{t} + \delta) - M(\bar{t}) \geq H(x^*(\bar{t} + \delta), \tau, \lambda(\bar{t} + \delta)) - H(x^*(\bar{t}), \tau, \lambda(\bar{t})).$$

Divide by δ and then let $\delta \to 0$. This shows

$$\begin{aligned} M'(\bar{t}) &\geq \left.\frac{d}{dt} H(x^*(t), \tau, \lambda(t))\right|_{t=\bar{t}} \\ &= H_x(x^*(\bar{t}), \tau, \lambda(\bar{t}))\, (x^*)'(\bar{t}) + H_\lambda(x^*(\bar{t}), \tau, \lambda(\bar{t}))\, \lambda'(\bar{t}) \\ &= -\lambda'(\bar{t})\, (x^*)'(\bar{t}) + (x^*)'(\bar{t})\, \lambda'(\bar{t}) = 0. \end{aligned}$$

By the same argument,

$$M(\bar{t}) - M(\bar{t} - \delta) \leq H(x^*(\bar{t}), \tau, \lambda(\bar{t})) - H(x^*(\bar{t} - \delta), \tau, \lambda(\bar{t} - \delta)).$$

Dividing by δ and letting $\delta \to 0$, we see $M'(\bar{t}) \leq 0$. Hence, $M' = 0$ almost everywhere. Combined with the fact that M is continuous, we see M is constant. □

The results given in Theorems 2.4 and 2.5 remain true in later chapters, even though the optimal control problems become more complicated (i.e., states fixed at the final time, bounds on the control, multiples states and controls). We do not reprove the results each time, as they follow in more or less the same manner. The proof for one of the more complicated cases is given in [158].

2.5 Exercises

Exercise 2.1 Complete the proof of Theorem 2.3 by proving the uniqueness statement.

Exercise 2.2 Reconsider Example 1.2,

$$\min_u \frac{1}{2} \int_{t_0}^{t_1} 3x(t)^2 + u(t)^2 \, dt$$

subject to $\quad x'(t) = x(t) + u(t), \ x(t_0) = x_0.$

For what values of t_0, t_1, and x_0 can we apply the Principle of Optimality to solve the problem using the solution to the original problem found in Chapter 1?

Exercise 2.3 Consider

$$\max_u \int_{t_0}^{t_1} (u(t)x(t) - u(t)^2 - x(t)^2) \, dt$$

subject to $\quad x'(t) = x(t) + u(t), \ x(t_0) = x_0.$

Note that the $t_0 = 1, t_1 = 5, x_0 = 2$ case is Exercise 1.3. Solve the problem for $t_0 = 1$, $t_1 = 3$, $x_0 = 2$ and note the results do not match the results from Exercise 1.3 on the interval $[1,3]$. For $t_0 = 3$ and $t_1 = 5$, what value of x_0 does the Principle of Optimality guarantee agreement with the solutions from Exercise 1.3 on $[3,5]$?

Chapter 3

State Conditions at the Final Time

Up to this point, we have viewed the value of the state at the terminal time to be immaterial, i.e., the objective functional (our goal) did not explicitly depend on $x(t_1)$. However, there are situations where we might wish to take it into consideration.

3.1 Payoff Terms

Many times, in addition to maximizing (or minimizing) terms over the entire time interval, we will wish to also maximize a function value at one particular point in time, specifically, the end of the time interval. For example, suppose you want to minimize the tumor cells at the final time in a cancer model, or the number of infected individuals at the final time in an epidemic model. The necessary conditions must be appropriately altered. In general, consider the following set-up,

$$\max_u \left[\phi(x(t_1)) + \int_{t_0}^{t_1} f(t, x(t), u(t))\, dt \right]$$

$$\text{subject to} \quad x' = g(t, x(t), u(t)), \; x(t_0) = x_0,$$

where $\phi(x(t_1))$ is a goal with respect to the final position or population level, $x(t_1)$. We call $\phi(x(t_1))$ a *payoff term*. It is sometimes referred to as the salvage term. Consider the resulting change in the derivation of the necessary conditions. Our objective functional becomes

$$J(u) = \int_{t_0}^{t_1} f(t, x(t), u(t))\, dt + \phi(x(t_1)).$$

In the calculation of

$$0 = \lim_{\epsilon \to 0} \frac{J(u^\epsilon) - J(u^*)}{\epsilon},$$

the only change occurs in the conditions at the final time,

$$0 = \int_{t_0}^{t_1} \left[(f_x + \lambda g_x + \lambda') \frac{dx^\epsilon}{d\epsilon} \bigg|_{\epsilon=0} + (f_u + \lambda g_u) h \right] dt \quad (3.1)$$
$$- \left(\lambda(t_1) - \phi'(x(t_1)) \right) \frac{\partial x^\epsilon}{\partial \epsilon}(t_1) \bigg|_{\epsilon=0}.$$

So, if we choose the adjoint variable λ to satisfy the previous adjoint equation and also

$$\lambda'(t) = -f_x(t, x^*, u^*) - \lambda(t) g_x(t, x^*, u^*),$$
$$\lambda(t_1) = \phi'(x^*(t_1)),$$

then (3.1) reduces to

$$0 = \int_{t_0}^{t_1} (f_u + \lambda g_u) h \, dt,$$

and the optimality condition

$$f_u(t, x^*, u^*) + \lambda g_u(t, x^*, u^*) = 0$$

follows as before. So, the only change in the necessary conditions is in the transversality condition

$$\lambda(t_1) = \phi'(x^*(t_1)).$$

To clarify how to calculate this adjoint final time condition, consider the following examples.

Example 3.1

$$\max_u \int_0^T f(t, x(t), u(t)) \, dt + 5x(T)^3$$
$$\text{subject to} \quad x'(t) = g(t, x(t), u(t)), \ x(0) = x_0.$$

Here we have

$$\phi(s) = 5s^3 \quad \Rightarrow \quad \phi'(s) = 15s^2,$$

so that the transversality condition is

$$\lambda(T) = 15x^*(T)^2.$$

Example 3.2 (from [100])

$$\min_u \frac{1}{2}\int_0^1 u(t)^2\,dt + x(1)^2$$
$$\text{subject to}\quad x'(t) = x(t) + u(t),\ x(0) = 1.$$

Note, this problem is identical to Example 1.1, except for the addition of the payoff term. So now, our goal includes minimizing the term $x(1)^2$, in addition to the square integral of the control. We can view this as minimizing a population, with exponential growth, at the end of a time frame. We should expect u to be negative, in order to decrease x, but $|u|$ cannot be too large because of the integral. In this example,

$$H = \frac{1}{2}u^2 + \lambda x + \lambda u.$$

The optimality condition gives

$$0 = \frac{\partial H}{\partial u} = u + \lambda \Rightarrow u^*(t) = -\lambda(t).$$

Also, the adjoint equation is

$$\lambda'(t) = -\frac{\partial H}{\partial x} = -\lambda \Rightarrow \lambda(t) = Ce^{-t},$$

for some constant C. Hence,

$$u^*(t) = -\lambda(t) = -Ce^{-t}.$$

So,

$$x'(t) = x - Ce^{-t},\ x(0) = 1,$$

which gives

$$x^*(t) = \frac{C}{2}e^{-t} + Ke^t,$$

where K is a constant. Recall, the transversality condition here is

$$\lambda(1) = \phi'(x(1)) = (x^2(1))' = 2x(1).$$

We have the system of linear equations

$$1 = x(0) = \frac{C}{2} + K$$
$$Ce^{-1} = \lambda(1) = 2x(1) = Ce^{-1} + 2Ke^1,$$

which can be solved to give $C = 2,\ K = 0$. Thus,

$$x^*(t) = e^{-t}, \ u^*(t) = -2e^{-t},$$

and u^* is negative as expected.

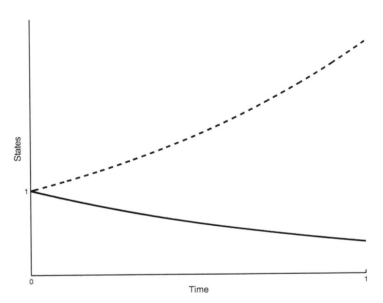

FIGURE 3.1: The optimal state for Example 3.2 (solid) and Example 1.1 (dashed).

If we plug these variables back into the objective functional, we find $J(u^*) = 1$. In the original Example 1.1, without the $x(1)^2$ term, $J(u^*) = 0$. The two optimal states are plotted in Figure 3.1.

Example 3.3 Let $x(t)$ represent the number of tumor cells at time t (with exponential growth factor α), and $u(t)$ the drug concentration. We wish to simultaneously minimize the number of tumor cells at the end of the treatment period and the accumulated harmful effects of the drug on the body. So, the problem is

$$\min_u \ x(T) + \int_0^T u(t)^2 \, dt$$

subject to $\quad x'(t) = \alpha x(t) - u(t), \ x(0) = x_0 > 0.$

This model is very simple and unrealistic; we use it for illustrative purposes only. A more sophisticated and interesting model is used in Lab 5.

Note that $\phi(s) = s$ here, so that $\phi'(s) = 1$. First, we construct the Hamiltonian and then calculate the necessary conditions:

$$H = u^2 + \lambda(\alpha x - u),$$

$$\frac{\partial H}{\partial u} = 2u - \lambda = 0 \text{ at } u^* \Rightarrow u^* = \frac{\lambda}{2},$$

$$\lambda' = -\frac{\partial H}{\partial x} = -\alpha\lambda \Rightarrow \lambda = Ce^{-\alpha t},$$

$$\lambda(T) = 1.$$

This gives the adjoint variable,

$$\lambda(t) = e^{\alpha(T-t)}.$$

Hence, we obtain the optimal control

$$u^*(t) = \frac{e^{\alpha(T-t)}}{2},$$

and we can then solve for the optimal state

$$x' = \alpha x - u = \alpha x - \frac{e^{\alpha(T-t)}}{2}, \quad x(0) = x_0.$$

This ODE can be solved using an integration factor to find

$$x^*(t) = x_0 e^{\alpha t} + e^{\alpha T}\frac{e^{-\alpha t} - e^{\alpha t}}{4\alpha}.$$

3.2 States with Fixed Endpoints

There are various possibilities of fixing the position of the state at the beginning or at the end of the time interval or both. The objective functional could depend on the final or initial position. Consider the problem

$$\max_u \int_{t_0}^{t_1} f(t, x(t), u(t))\, dt + \phi(x(t_0))$$

subject to $\quad x'(t) = g(t, x(t), u(t)),$

$\qquad\qquad\quad x(t_0)$ free, $x(t_1) = x_1$ fixed.

This is different than the problems we have been examining, as the state is fixed at the end of the time interval, not at the beginning. However, the same argument we used in Chapter 1, with the adjoint chosen appropriately, shows that the necessary conditions for an optimal pair u^*, x^* will be as before, with only the transversality condition changed. Specifically,

$$\lambda(t_0) = \phi'(x(t_0)).$$

See Exercise 3.4. This suggests there may exist a simple duality between the state and adjoint boundary conditions.

Consider the problem below, where the state is fixed at both the beginning and end of the time interval,

$$\max_{u} \int_{t_0}^{t_1} f(t, x(t), u(t))\, dt$$

subject to $\quad x'(t) = g(t, x(t), u(t))$
$\qquad\qquad\ x(t_0) = x_0,\ x(t_1) = x_1\ \text{both fixed}.$ (3.2)

The maximization here is over all *admissible* controls. That is, the set of controls which adhere to all stated restrictions (explicit and implicit). In the case of (3.2), this would mean all controls which steer the state from the fixed initial condition to the fixed final condition. A slight modification of the necessary conditions is needed to solve such a problem. We give the following theorem.

THEOREM 3.1
If $u^*(t)$ and $x^*(t)$ are optimal for problem (3.2), then there exists a piecewise differentiable adjoint variable $\lambda(t)$ and a constant λ_0, equal to either 0 or 1, such that

$$H(t, x^*(t), u(t), \lambda(t)) \leq H(t, x^*(t), u^*(t), \lambda(t))$$

for all admissible controls u at each time t, where the Hamiltonian H is

$$H = \lambda_0 f(t, x(t), u(t)) + \lambda(t) g(t, x(t), u(t))$$

and

$$\lambda'(t) = -\frac{\partial H(t, x^*(t), u^*(t))}{\partial x}.$$

The proof of this result is somewhat different from the proof technique we have used for necessary conditions so far. It is also more difficult, as the state is now overdetermined, i.e., a first order ODE with two boundary conditions. Only controls yielding the required state boundary conditions can be considered. So, we refrain from giving the proof here. For more information see [62, 100]. Note, as x now has both boundary conditions, λ has none.

The constant λ_0 arises here to adjust for degenerate problems, or problems where the objective functional is immaterial. Namely, all admissible controls yield the same objective functional value. To motivate this, recall the original proof of the necessary conditions given in Chapter 1. We start with an optimal control and form a family of controls $u^* + \epsilon h$. We cannot do something this simple minded here, as any control must also satisfy the constraints on the state. We can use a similar technique (albeit more complicated) so long as we have an optimal control and a different control which yield distinct objective functional values. If this cannot be done, we are forced into the $\lambda_0 = 0$ case. Consider the following example.

Example 3.4 (from [55])

$$\min_u \int_0^1 u(t)\, dt$$

subject to $x'(t) = u(t)^2$, $x(0) = 0$, $x(1) = 0$.

If we examine the differential equation, it is clear that $u \equiv 0$ is the only control which produces a state x satisfying the boundary conditions. Therefore, it is automatically the optimal control. However, let us examine the necessary conditions. First, suppose we are in the $\lambda_0 = 1$ case. Then,

$$H = u + u^2 \lambda,$$
$$\lambda' = -\frac{\partial H}{\partial x} = 0.$$

This shows that $\lambda \equiv c$ for some constant c. Now, the condition that H is maximized at u^* gives the familiar condition $0 = \frac{\partial H}{\partial u}$. Namely,

$$0 = 1 + 2\lambda u^* = 1 + 2cu^* \Rightarrow u^* \equiv -1/2c.$$

Hence, $x' = 1/4c^2$. But, this is incompatible with the boundary conditions. Thus, $\lambda_0 \neq 1$.

On the other hand, it is easily checked that $u = 0$ satisfies all conditions of Theorem 3.1 when $\lambda_0 = 0$.

Of course, in application, objective functionals which are immaterial would not be chosen. As all further problems presented here will be nondegenerate, we assume $\lambda_0 = 1$. In this case, the conditions of Theorem 3.1 appear similar to the necessary conditions we have seen so far, except the terminal boundary condition is now on x. The following examples illustrate how to solve such problems.

Example 3.5

$$\min_u \int_0^4 u(t)^2 + x(t)\,dt$$

subject to $x'(t) = u(t)$, $x(0) = 0$, $x(4) = 1$.

We begin by forming the Hamiltonian

$$H = u^2 + x + \lambda u.$$

We have no transversality condition, as x has both boundary conditions, but we make use of the adjoint condition,

$$\lambda'(t) = -\frac{\partial H}{\partial x} = -1 \Rightarrow \lambda(t) = k - t$$

for some constant k. Then, the optimality condition gives

$$0 = \frac{\partial H}{\partial u} = 2u + \lambda \quad \Rightarrow \quad u^* = -\frac{\lambda}{2} = \frac{t-k}{2}.$$

Solving the state equation with this control gives

$$x^*(t) = \frac{t^2}{4} - \frac{kt}{2} + c$$

for some constant c. Using the boundary conditions, $x(0) = 0$ implies $c = 0$, and $x(4) = 1$ gives $k = \frac{3}{2}$. So,

$$u^*(t) = \frac{2t-3}{4} \quad \text{and} \quad x^*(t) = \frac{t^2 - 3t}{4}.$$

Example 3.6 (from [100])

$$\min_u \frac{1}{2}\int_0^1 u^2(t)\,dt$$

subject to $x'(t) = x(t) + u(t)$, $x(0) = 1$, $x(1) = 0$.

This is another variation on Examples 1.1 and 3.2. The objective functional once again does not depend on x, but we must choose a control that moves x from 1 to 0. Again, we expect a negative u. The Hamiltonian is

$$\frac{1}{2}u^2 + \lambda x + \lambda u.$$

As before, the optimality condition gives

$$0 = \frac{\partial H}{\partial u} = u + \lambda \Rightarrow u^* = -\lambda.$$

Also,

$$\lambda'(t) = -\frac{\partial H}{\partial x} = -\lambda \Rightarrow \lambda(t) = Ce^{-t},$$

for some constant C. Thus,

$$u^*(t) = -\lambda(t) = -Ce^{-t},$$

so that

$$x'(t) = x - Ce^{-t} \Rightarrow x^*(t) = \frac{C}{2}e^{-t} + Ke^t.$$

Enforcing the boundary conditions on x, we find

$$1 = x(0) = \frac{C}{2} + K$$
$$0 = x(1) = \frac{C}{2}e^{-1} + Ke,$$

which gives $C = \frac{2e^2}{e^2-1}$ and $K = \frac{1}{1-e^2}$, so that

$$x^*(t) = \frac{1}{e^2-1}(e^{2-t} - e^t), \quad u^*(t) = \frac{2}{1-e^2}e^{2-t}.$$

Note, in Example 3.2 we wanted to minimize the value of $x(1)^2$ and the cumulative effect of the control. So, we wanted to push $x(1)$ close to 0. Here, we choose the control with the smallest cumulative effect that forces the state to 0. If we plug the optimal control from this example into the objective functional, we find $J(u^*) = 2(1 - e^{-2})^{-1}$, whereas the value of $J(u^*)$ in Example 3.2 was 1. Not fixing the final state allows more freedom in the choice of controls, and the objective functional can be reduced further. The two optimal states are shown in Figure 3.2.

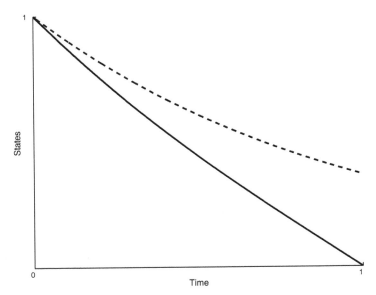

FIGURE 3.2: The optimal state from Example 3.6 (solid) is forced to 0. The optimal state from Example 3.2 (dashed) is not.

3.3 Exercises

Exercise 3.1 Solve

$$\min_u \frac{1}{2} \int_0^1 u(t)^2 \, dt + x(1)$$

subject to $\quad x'(t) = -u(t), \ x(0) = 1.$

Exercise 3.2 (from [169]) Let $d, S, x_0, T > 0$ be positive constants. Solve

$$\max_u \int_0^T \left(x(t) - \frac{1}{2} u(t)^2 \right) dt + Sx(T)$$

subject to $\quad x'(t) = -dx(t) + u(t), \ x(0) = x_0.$

Exercise 3.3 Solve

$$\min_u \frac{1}{2}\int_a^b u(t)^2\, dt$$

subject to $x'(t) = u(t) - x(t),\ x(a) = 1,\ x(b) = 2.$

Exercise 3.4 State and prove the necessary conditions for the problem

$$\max_u \int_{t_0}^{t_1} f(t, x(t), u(t))\, dt + \phi(x(t_0))$$

subject to $x'(t) = g(t, x(t), u(t)),\ x(t_1) = x_1.$

Exercise 3.5 (from [169]) Consider the problem

$$\min_u \frac{1}{4}\int_0^1 u(t)^4\, dt$$

subject to $x'(t) = x(t) + u(t),\ x(0) = x_0,\ x(1) = 0.$

Show that the optimal control is $u^*(t) = \dfrac{4x_0}{3(e^{-4/3} - 1)} e^{-t/3}.$

Exercise 3.6 (from [100]) Show that there can be no optimal control for

$$\max_u \int_0^1 u(t)\, dt$$

subject to $x'(t) = x(t) + u(t)^2,\ x(0) = 1,\ x(1) = 0.$

Exercise 3.7 (from [126]) Optimal control techniques can be used to verify the shortest distance between two points is a straight line. If we have two points in $\mathbb{R}^2$, we can rescale so that one point is $(0,0)$. Let the other point (after rescaling) be (a, b). Let the state $x(t)$ be a curve from $(0,0)$ to (a, b). We take the control to be $u(t) = x'(t)$. We are interested in minimizing arc length, which has the formula $\int_0^a \sqrt{1 + x'(t)^2}\, dt$. Therefore, our optimal control problem is

$$\min_u \int_0^a \sqrt{1 + u(t)^2}\, dt$$

subject to $x'(t) = u(t),\ x(0) = 0,\ x(a) = b.$

Show the optimal state is the straight line between $(0,0)$ and (a,b).

Exercise 3.8 (from [55]) Using the same technique as in the previous exercise, find the curve from $(0,2)$ to $(2,4)$ which, when revolved around the x-axis, has minimal surface area. The surface area of the revolution of the curve $x(t)$ (defined on $a \leq t \leq b$) is given by the formula $A = 2\pi \int_a^b x(t)\sqrt{1 + x'(t)^2}\, dt$.

Exercise 3.9 (from [126]) We wish to heat a room to a desired temperature D in a fixed time frame $[0, T]$. Let $x(t)$ be the temperature in the room and $u(t)$ the rate of heat supply. The temperature is governed by $x'(t) = -a(x(t) - D) + bu(t)$. If the initial temperature is 0 degrees ($x(0) = 0$), find the heating schedule $u(t)$ which reaches the desired temperature ($x(T) = D$) while minimizing energy used. Here the performance index is $\frac{1}{2}\int_0^T u(t)^2\, dt$.

Chapter 4

Forward-Backward Sweep Method

Consider the optimal control problem

$$\max_u \int_{t_0}^{t_1} f(t, x(t), u(t))\, dt$$

$$\text{subject to} \quad x'(t) = g(t, x(t), u(t)), \; x(t_0) = a.$$

We want to solve such problems numerically, that is, devise an algorithm that generates an approximation to an optimal piecewise continuous control u^*. We break the time interval $[t_0, t_1]$ into pieces with specific points of interest $t_0 = b_1, b_2, \ldots, b_N, b_{N+1} = t_1$; these points will usually be equally spaced. The approximation will be a vector $\vec{u} = (u_1, u_2, \ldots, u_{N+1})$, where $u_i \approx u(b_i)$.

There are various methods of this type which can be employed to solve optimal control problems. For example, total-enumeration methods or linear programming techniques can be employed [17]. However, as we saw in the previous chapters, any solution to the above optimal control problem must also satisfy

$$x'(t) = g(t, x(t), u(t)), \; x(t_0) = a,$$

$$\lambda'(t) = -\frac{\partial H}{\partial x} = -(f_x(t, x, u) + \lambda(t) g_x(t, x, u)), \; \lambda(t_1) = 0,$$

$$0 = \frac{\partial H}{\partial u} = f_u(t, x, u) + \lambda(t) g_u(t, x, u) \text{ at } u^*.$$

The third equation, the optimality condition, can usually be manipulated to find a representation of u^* in terms of t, x, and λ. If this representation is substituted back into the ODEs for x, λ, then the first two equations form a two-point boundary value problem. There exist many numerical methods to solve initial value problems, such as Runge-Kutta or adaptive schemes, and boundary value problems, such as shooting methods [22, 32]. Any of these methods could be used to solve the optimality system, and thus, the optimal control problem (if appropriate existence and uniqueness results are established).

We wish to take advantage of certain characteristics of the optimality system, however. First, we are given an initial condition for the state x but a final time condition for the adjoint λ. Second, g is a function of t, x, and u

only. Values for λ are not needed to solve the differential equation for x using a standard ODE solver. Taking this into account, the method we present here is very intuitive. It is generally referred to as the Forward-Backward Sweep method. Information about convergence and stability of this method can be found in [74]. A rough outline of the algorithm is given below. Here, $\vec{x} = (x_1, \ldots, x_{N+1})$ and $\vec{\lambda} = (\lambda_1, \ldots, \lambda_{N+1})$ are the vector approximations for the state and adjoint.

Step 1. Make an initial guess for $\vec{u}$ over the interval.

Step 2. Using the initial condition $x_1 = x(t_0) = a$ and the values for $\vec{u}$, solve $\vec{x}$ forward in time according to its differential equation in the optimality system.

Step 3. Using the transversality condition $\lambda_{N+1} = \lambda(t_1) = 0$ and the values for $\vec{u}$ and $\vec{x}$, solve $\vec{\lambda}$ backward in time according to its differential equation in the optimality system.

Step 4. Update $\vec{u}$ by entering the new $\vec{x}$ and $\vec{\lambda}$ values into the characterization of the optimal control.

Step 5. Check convergence. If values of the variables in this iteration and the last iteration are negligibly close, output the current values as solutions. If values are not close, return to Step 2.

An example of successive control estimates is shown in Figure 4.1. We make a few notes about the algorithm. For the initial guess, $\vec{u} \equiv 0$ is almost always sufficient. In certain problems, where division by u occurs for example, a different initial guess must be used. Occasionally, the initial guess may require adjusting if the algorithm has problems converging. Often in Step 4, it is necessary to use a convex combination between the previous control values and values given by the current characterization. This often helps to speed the convergence. As you will see, this is done in the provided codes. For Steps 2 and 3, any standard ODE solver can be used. For the purposes of this text, a Runge-Kutta 4 routine is used. Specifically, given a step size h and an ODE $x'(t) = f(t, x(t))$, the approximation of $x(t+h)$ given $x(t)$ is

$$x(t+h) \approx x(t) + \frac{h}{6}(k_1 + 2k_2 + 2k_3 + k_4) \tag{4.1}$$

where

$$\begin{aligned} k_1 &= f(t, x(t)) \\ k_2 &= f(t + \frac{h}{2}, x(t) + \frac{h}{2} k_1) \\ k_3 &= f(t + \frac{h}{2}, x(t) + \frac{h}{2} k_2) \\ k_4 &= f(t + h, x(t) + h k_3). \end{aligned} \tag{4.2}$$

Forward-Backward Sweep Method

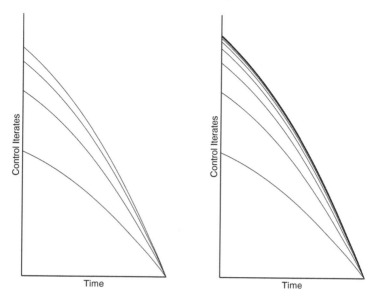

FIGURE 4.1: Control estimates are plotted. The first four iterations (after the initial guess) are plotted in the first graph, and the first fifteen in the second. Note the graphs are converging to the correct control.

The error for Runge-Kutta 4 is $\mathcal{O}(h^4)$. More information on the stability and accuracy of this and other Runge-Kutta routines is found in numerous texts. One of the classic references for these methods is Butcher [23, 24].

Many types of convergence tests exist for Step 5. Often times, it is sufficient to require $\|u - \vec{old u}\| = \sum_{i=1}^{N+1} |u_i - old u_i|$ to be small, where $\vec{u}$ is the vector of estimated values of the control during the current iteration, and $\vec{old u}$ is the vector of estimated values from the previous iteration. Here, $\|\cdot\|$ refers to the ℓ^1 norm for vectors, i.e., the sum of the absolute value of the terms. Both these vectors are of length $N+1$, as there are N time steps. In this text, we use a slightly stricter convergence test. Namely, we will require the relative error to be negligibly small, i.e.,

$$\frac{\|\vec{u} - \vec{old u}\|}{\|\vec{u}\|} \leq \delta$$

where δ is the accepted tolerance. We must make one small adjustment; we must allow for zero controls. So, multiply both sides by $\|\vec{u}\|$ to remove it from the denominator. Therefore, our requirement is

$$\delta \|\vec{u}\| - \|\vec{u} - \vec{old u}\| \geq 0,$$

or

$$\delta \sum_{i=1}^{N+1} |u_i| - \sum_{i=1}^{N+1} |u_i - oldu_i| \geq 0. \tag{4.3}$$

We will actually make this requirement of all variables, not just the control. In the lab programs, we take $N = 1000$ and $\delta = 0.001$.

The remainder of this chapter will be devoted to further explanation of the Forward-Backward Sweep algorithm by way of example.

Example 4.1

$$\max_u \int_0^1 Ax(t) - Bu(t)^2 \, dt$$

subject to $\quad x'(t) = -\dfrac{1}{2}x(t)^2 + Cu(t), \ x(0) = x_0 > -2,$

$\qquad\qquad\quad A \geq 0, B > 0.$

We require $B > 0$ so that this is a maximization problem. Before writing the code, we develop the optimality system of this problem by first noting the Hamiltonian is

$$H = Ax - Bu^2 - \frac{1}{2}\lambda x^2 + C\lambda u.$$

Using the optimality condition,

$$0 = \frac{\partial H}{\partial u} = -2Bu + C\lambda \Rightarrow u^* = \frac{C\lambda}{2B}.$$

We also easily calculate the adjoint equation to find

$$x'(t) = -\frac{1}{2}x^2 + Cu, \ x(0) = x_0$$
$$\lambda'(t) = -A + x\lambda, \ \lambda(1) = 0.$$

Using these two differential equations and the representation of u^*, we generate the numerical code as described above, written in MATLAB [138]. The code can be viewed in its entirety in the file *code1.m*, and is also shown in increments below.

Forward-Backward Sweep Method

```
─────────────── code1.m ───────────────
1  function y = code1(A,B,C,x0)
2
3  test = -1;
4
5  delta = 0.001;
6  N = 1000;
7  t = linspace(0,1,N+1);
8  h = 1/N;
9  h2 = h/2;
10
11 u = zeros(1,N+1);
12
13 x = zeros(1,N+1);
14 x(1) = x0;
15 lambda = zeros(1,N+1);
16
17 while(test < 0)
18
```

Line 1 establishes the MATLAB function code1 and variables A, B, C, and x_0 as inputs. The variable y is the output. The variable *test* created in Line 3 is the convergence test variable. It begins the *while* loop in Line 17. The loop, as we will see, contains the forward-backward sweep. Once convergence occurs, *test* will become non-negative, and the *while* loop will end. In Line 7, a vector $\vec{t}$ representing the time variable is created. The MATLAB function *linspace* creates $N+1 = 1001$ equally spaced nodes between 0 and 1, including 0 and 1. In Line 8, the spacing between these nodes is assigned as h. Line 9 establishes a convenient short-hand which is used in the Runge-Kutta subroutine. Line 11 is our initial guess for $\vec{u}$, namely, $u_i = 0$ at each of the 1001 nodes. Lines 13 and 15 declare the vectors $\vec{x}$ and $\vec{\lambda}$ and their size. These are not guesses, as these values will be overwritten during the sweep process. The initial value of $\vec{x}$ is stored in Line 14.

```
─────────────── code1.m ───────────────
19      oldu = u;
20      oldx = x;
21      oldlambda = lambda;
22
```

Lines 19 - 21 are the first lines inside the *while* loop, which begins the sweep process. These lines store the vectors $\vec{u}$, $\vec{x}$, and $\vec{\lambda}$ as previous values, denoted as $\vec{oldu}$, $\vec{oldx}$, and $\vec{old\lambda}$. Recall that our convergence test requires the values of the current and previous iterations. After storing the current values as the previous ones here, new values are generated in the following lines.

```
for i = 1:N
    k1 = -0.5*x(i)^2 + C*u(i);
    k2 = -0.5*(x(i) + h2*k1)^2 + C*0.5*(u(i) + u(i+1));
    k3 = -0.5*(x(i) + h2*k2)^2 + C*0.5*(u(i) + u(i+1));
    k4 = -0.5*(x(i) + h*k3)^2 + C*u(i+1);
    x(i+1) = x(i) + (h/6)*(k1 + 2*k2 + 2*k3 + k4);
end
```

Lines 23 - 29 contain the Runge-Kutta sweep solving $\vec{x}$ forward in time. Line 23 begins the *for* loop and Line 29 ends it. Line 24 calculates the k_1 value, which is simply the RHS of the differential equation. In Line 25, to find k_2, x is replaced with $x + \frac{h}{2}k_1$. We are also to adjust the time variable t by replacing it with $t + \frac{h}{2}$. There is no explicit dependence on t in the differential equation, but u is a function of t. So in calculating k_2 and k_3, we should replace u_i with $u_{i+h/2}$. However, this value is not assigned by our vector, meaning there is no u component halfway between i and $i+1$ locations. There are many ways to approximate this value. An interpolating polynomial or spline of u could be generated, for example. However, it usually suffices to approximate it with the average

$$\frac{u(i) + u(i+1)}{2}.$$

In Line 26, similar prescribed changes in x and u are made. In Line 27, a full time step is called for, so u_{i+1} is used. Line 27 generates the next iterated value of the state x. Note, as x_1 is used to find x_2, x_2 to find x_3, and so on, x_N is used to find x_{N+1}. This is why in Line 23, i only runs to N, not $N+1$.

```
for i = 1:N
    j = N + 2 - i;
    k1 = -A + lambda(j)*x(j);
    k2 = -A + (lambda(j) - h2*k1)*0.5*(x(j)+x(j-1));
    k3 = -A + (lambda(j) - h2*k2)*0.5*(x(j)+x(j-1));
    k4 = -A + (lambda(j) - h*k3)*x(j-1);
    lambda(j-1) = lambda(j) - ...
        (h/6)*(k1 + 2*k2 + 2*k3 + k4);
end
```

Lines 31 - 39 consist of the Runge-Kutta sweep solving $\vec{\lambda}$ backward in time. The *for* loop begins in Line 30, while the new index is introduced in Line 32. Notice that as i counts forward from 1 to N, j counts backward from $N+1$ to 2. Line 33 is the k_1 calculation, which comes directly from the differential equation. In Line 34, λ_j is replaced by $\lambda_j - \frac{h}{2}k_1$. Notice the minus sign, as we are moving backward in time, so the time increment should actually be $-1/N$. As before, we approximate a backward half time-step of $\vec{x}$ by an average. Lines 35 and 36 are the remaining two steps. Lines 37 and 38 are the approximation step. Note the use of the ellipse ... in the code on line 37. This tells MATLAB that 37 and 38 are really the same line of code, and to treat the line break only as a regular space. MATLAB does not require this, nor does it care how long each line of code is. It is done here for printing purposes. Line 39 ends the *for* loop. Note, each λ_i value is used to find the one before it, so that λ_2 is used to find λ_1. This is why we only need to count backward to 2, not 1.

```
                            code1.m
41      u1 = C*lambda/(2*B);
42      u = 0.5*(u1 + oldu);
43
44      temp1 = delta*sum(abs(u)) - sum(abs(oldu - u));
45      temp2 = delta*sum(abs(x)) - sum(abs(oldx - x));
46      temp3 = delta*sum(abs(lambda)) - ...
47          sum(abs(oldlambda - lambda));
48      test = min(temp1, min(temp2, temp3));
49  end
50
```

Line 41 is the representation of $\vec{u}$ using the new values for $\vec{\lambda}$. This is not stored as the control $\vec{u}$, but as a temporary vector $\vec{u1}$. The control $\vec{u}$ is set as the average of the last iteration of $\vec{u}$, namely $\vec{oldu}$, and the new representation. This is the convex combination described earlier. Lines 44, 45, and 46/47 are the convergence test parameters of each variable, where $\delta = 0.001$. Recall, we require these three values to be non-negative. In Line 48, the variable *test* is reassigned as the minimum of these three values. The MATLAB function *min* is a binary operation, so in order to find the minimum of three values, two applications of *min* are necessary. The *end* in Line 49 marks the end of the *while* loop. If the minimum is non-negative, i.e., if all three values are non-negative, then convergence has been achieved, as per (4.3), and the *while* loop ends. If it is not, then the program returns to the beginning of the *while* loop and performs another sweep. Once convergence occurs, the values of the final vectors are stored in the output matrix y.

```
                            code1.m
51  y(1,:) = t;
52  y(2,:) = x;
53  y(3,:) = lambda;
54  y(4,:) = u;
```

Note, this technique can only be used to solve problems where the state is fixed at the initial time and free at the terminal time. An obvious adjustment to the code allows states fixed at the terminal time and free at the initial time. However, different algorithms, specifically ones based on shooting methods, must be employed if the state has two boundary conditions. One such solver is discussed in Chapter 21. For more information on other solvers, see [6, 7, 139].

We also would like to point out different techniques for control updates. The term convex combination may have been confusing, as only a simple average was taken between the previous and current control estimates. Convex combination refers to an entire family of updating procedures. Many of these arise in the bounded control case, which we will discuss in Chapter 8. A different kind of convex combination which could have been used here is

$$u1 * (1 - c^k) + oldu * c^k,$$

where k is the current iteration and $0 < c < 1$. The terms $u1$ and *oldu* are just as in the code, the control estimate from the characterization and the previous sweep, respectively. This is a weighted average, where the weight shifts each iteration towards the current iteration.

Approaches such as these have benefits and drawbacks. When compared to our convex combination, the simple average, this new method will often converge more quickly, i.e., with less iterations. However, this usually leads to loss of accuracy. Often times, methods like this converge "too quickly," stopping the sweep process before our code would. On the other hand, these methods have proven to be useful alternatives for problems where simpler approaches failed to converge. Throughout this book, we will use the simpler approach, because it works. However, the reader should keep in mind that more complicated techniques can be used when necessary.

Chapter 5

Lab 1: Introductory Example

We now begin working on the first few interactive lab programs. They will allow you to experiment with optimal control problems and see the solutions. Most of the labs are based on current applied mathematical research, dealing with an array of biological problems. The first is the problem from the preceding chapter, and the code used is exactly what we developed there. Before preceding, however, we need to clarify a few details about the programs and MATLAB.

First, while MATLAB is needed to run the provided programs, it is certainly not needed to solve optimal control problems in general. Any mathematical programming language, such as FORTRAN or C++, is capable of the calculations needed. MATLAB was chosen for this text because, in the opinion of the authors, it is easily accessible and has superior graphing tools.

On that note, however, the programs used in this workbook are designed so that no knowledge of MATLAB is required. For each problem, there is a user-friendly interface that will guide you through. Each lab consists of two different MATLAB programs, *lab_.m* and *code_.m*. For example, there are two programs associated with Lab 1, *lab1.m* and *code1.m*. Here, *.m is the extension given to all files intended for use in MATLAB. The file *code1.m* is the Runge-Kutta based, forward-backward sweep solver we built in the previous chapter. It takes as input the values of the various parameters in the problem and outputs the solution to the optimality system. The file *lab1.m* is the user-friendly interface. It will ask you to enter the values of the parameters one by one, compile *code1.m* with these values, and plot the resulting solutions. All the files must be in the directory that MATLAB treats as the home directory. This is usually the *Work* directory.

If you have experience with MATLAB, you may wish to not use the interface and instead use only the actual codes. They operate as standard MATLAB function files, with the parameters entered as input. This will allow you a little more freedom than the interface. However, the interface, especially when going through the labs, is very convenient and will most likely save time. If you do choose to use only the *code_* files, you will need to run the interface a few times before starting the labs in order to see exactly what they do, so that you can emulate them on your own.

If you are not a seasoned MATLAB veteran, do not worry. This book is written with you in mind. However, we do need to cover just a few basic things about MATLAB. When you open MATLAB, there will be several different windows or portals. There will be one, most likely the largest and most likely on the right side, called the Command Window. Everything we do in this book will take place here. In the Command Window, there will be a prompt. This is where you will type your commands. For example, to open the interface for Lab 1, simply type *lab1* at the prompt and press enter.

One of the more important commands to know is the stop command. Any time you wish MATLAB to stop what it is doing, simply hit *Ctrl-c*. This will kill the current application and return the prompt. It will also report to you exactly what it was doing when you gave the stop order, but this will rarely be of interest to you. The command *Ctrl-c* may be useful when you enter certain parameters. Ill-conditioned problems or problems with invalid parameter values will not necessarily converge. This will not stop MATLAB from trying, however. It will continue to sweep forward and backward until it is stopped. All the data provided in the labs is taken from the research, so convergence always occurs. However, when you supply your own data, which you are highly encouraged to do, you have no such guarantee. Unless otherwise specified in the lab, convergence should take no longer than 30 seconds. If it has failed to do so by then, stop the application and try different numbers.

The interface should be self-explanatory. Once opened, it will ask you to enter a value for the first parameter. Type a number and press enter. If you fail to type a number, or enter a number which is not of the right type, you will receive an error message and be asked for the parameter again. If you accidentally enter the wrong value and press enter, simply hit *Crtl-c* and begin the *lab_* program again. Once all the parameters are entered, it will display "One moment please ..." as it compiles the solutions using the *code_* program. After the solutions have been found, it will ask if you would like to vary any parameters. If you respond negatively, it will automatically plot the optimal solutions in labeled graphs. The graphs may appear too small to view. However, if you expand the window, the graphs will enlarge appropriately. If you reply positively about varying the parameters, it will ask which parameter to vary. You will then be asked to enter a second value for this parameter. It will compile a second set of solutions. This set represents solutions for the problem with the same parameters as before, except the chosen parameter will be changed to the new value. Then, both solutions will be plotted together, with the original solutions plotted in blue, and the second set plotted in green. This will allow you to evaluate how each variable affects the optimal system. When you are done studying the graph, simply go back to the prompt and retype *lab_* to start the interface over.

Finally, you are encouraged to compute the optimality system for each problem by hand. Then, open the *code_.m* files to see that system translated to MATLAB code. All the programs work more or less the same way and are written in a uniform manner that should make them easy to read. To

Lab 1: Introductory Example

open an m-file, select the OPEN option under FILE in the upper left-hand corner. There may also be a standard open folder icon at the top of your screen, depending on how your MATLAB interface is organized. This will open a window showing all the m-files in your *Work* folder.

If you have any questions about MATLAB, you can refer to the MATLAB manual or to any of the numerous MATLAB guides available. However, the MATLAB Help menu within the program is one of the best resources of information.

This first lab will utilize the code developed in Chapter 4 in order to solve the following optimal control problem.

$$\max_u \int_0^1 Ax(t) - Bu^2(t)\, dt$$
$$\text{subject to} \quad x'(t) = -\frac{1}{2}x^2(t) + Cu(t),\ x(0) = x_0 > -2,$$
$$A \geq 0, B > 0.$$

To begin the program, open MATLAB. At the prompt, type *lab1* and press enter. To become acquainted with the program, perform a few test runs. Enter values for the constants A, B, C, and x_0. At first, do not vary any parameters. The graphs of the resulting optimal solutions, i.e., the adjoint and the optimal control and state, will automatically appear. Run the program again, enter different values, and vary one of the parameters. Once you feel comfortable with the structure of the program, begin working through the lab exercises below.

This lab will focus on using the program to characterize the optimal control and resulting state and to ascertain how each parameter affects the solution. First, let us consider the goal of the problem. On one hand, we want to use the control u to maximize the integral of x. On the other hand, we also want to maximize the negative squared value of u. This, of course, is equivalent to minimizing the squared value of u. Thus, we must find the right balance of increasing x and keeping u as small as possible. Enter the values

$$\boxed{A = 1 \quad B = 1 \quad C = 4 \quad x_0 = 1} \qquad (5.1)$$

and do not vary any parameters, then look at the solutions. Your output should look something like Figure 5.1. We see u begins strongly, pushing x up but steadily decreasing to 0. This makes logical sense when we consider the differential equation of x. Undisturbed by u, the state x will decrease monotonically. So, we want to push x up early in the time period, so that the natural decay will be less significant. As we only care about minimizing the integral of u, and the distribution is irrelevant, the control should be highest early on. We see this is exactly what the optimal control is. Also, note that x begins to decrease at the end of the interval, as the control approaches zero.

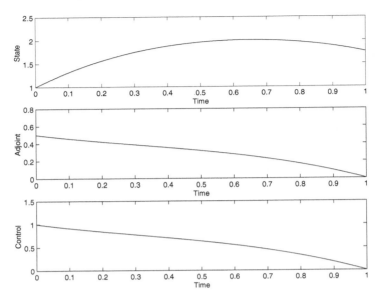

FIGURE 5.1: The optimal state, adjoint, and control for the values (5.1).

Reenter the values in (5.1) and then vary the initial condition with $x_0 = 2$. As the second state begins higher, less control is needed to achieve a similar effect. Notice that the second control begins lower than the first, but they quickly approach each other and are almost identical by $t = 0.6$. This causes the two states to move towards each other as well, although they never actually meet. Now use $x_0 = -1$. This time, x begins below zero, so a greater control is needed to push the state up more quickly. Notice, however, we see the same effect as before, where the two controls eventually merge, although, much later than in the previous simulation. We mention here why the requirement $x_0 > -2$ is imposed. If you were to solve the state equation without u (i.e., $C = 0$), you would find $x_0 > -2$ is required, or division by 0 will occur and the state will blow-up in finite time. However, we know u will be used to increase x, so this condition is sufficient to give a finite state solution with the control.

Use the (5.1) values, varying C with $C = 1$. We have decreased the effect u has on the growth of the state. The optimal control in the second system is less than in the first. It is worth using a greater control in the first system, as it is more effective. Also, the second state, unlike the others we have seen, is decreasing over the whole interval. What little control is used does not increase the state, but only neutralizes some of the natural decay. It would now take far too much control to increase the state. Enter the same parameter values, this time varying with $C = 8$. The results are as you might expect. The second optimal control, now more effective, is greater than the first. The second state increases far more than the first, but still decreases as its control approaches zero. Finally, note that when C is varied, we do not have the two controls merging together.

Enter (5.1) and vary with $C = -4$. The control now has the opposite effect on the growth of the state. We see the control for the second state is merely the first control reflected across the x-axis, while the state and adjoint are the same. Try $C = 0$. Here, the control has no effect on x, so the optimal control is $u \equiv 0$, regardless of A, B, or x_0.

Reenter (5.1). Choose to vary A. Specifically, try $A = 4$ as your second value. In the second system, $A = 4B$, so maximizing $x(t)$ is four times as important as minimizing u^2. We see this playing out in the solutions. A greater u is used so that x can be increased appropriately. Conversely, enter (5.1) varying with $B = 4$. In this case, minimizing $u(t)^2$ is more important. We see on the graph, $u(t)$ is pulled closer to zero, even though this causes $x(t)$ to increase much less at the beginning. The constants A and B are called *weight parameters*, as they determine the importance or weight of variables in the objective functional.

If you were to compare the graphs of the optimal solutions for

$$\boxed{A = 1 \quad B = 2 \quad C = 4 \quad x_0 = 1} \tag{5.2}$$

to the solutions for

$$\boxed{A = 2 \quad B = 4 \quad C = 4 \quad x_0 = 1} \tag{5.3}$$

you would notice they were exactly the same. This is because the system is only influenced by the ratio of the constants A and B, not the actual values. We know $B \neq 0$, so we could divide it out of the integral. This would make our objective function

$$B \int_0^1 \frac{A}{B} x(t) - u(t)^2 \, dt.$$

Of course, the constant B in front of the integral is irrelevant, so we ignore it. Thus, the only constant of significance in the integrand is $\frac{A}{B}$. In all future labs, one term of the integrand will have no weight parameter, as it has been divided out.

Before finishing, we look at a few special cases. Try $A = 0$. This will also cause the trivial solution $u^* \equiv 0$ regardless of B, C, and x_0. If we no longer care about maximizing x, then we clearly should simply pick $u \equiv 0$ and ignore x. We cannot choose $B = 0$, because we divide by B in the characterization of the control. However, a similar situation occurs as we let $B \to 0$. For instance, try $A = 1$ and $B = 0.01$. Then, compare the graphs to $A = 1$ and $B = 0.00001$. A very large u (or large negative u, if $C < 0$) is used to push x

up as quickly as possibly, because almost no importance is placed on keeping u^2 small.

Exercise 5.1 Reconsider the problem with $B = 0$

$$\max_u \int_0^1 Ax(t)\, dt$$

subject to $x'(t) = -\frac{1}{2}x^2(t) + Cu(t),\ x(0) = x_0 > -2,$

$A \geq 0.$

Show (analytically) that no optimal control can exist when $A > 0$.

Chapter 6

Lab 2: Mold and Fungicide

For the second lab, we will explore an optimal control problem with biological applications. Let $x(t)$ be a population concentration at time t, and suppose we wish to reduce the population over a fixed time period. We will assume x has a growth rate r and carrying capacity M. The application of a substance is known to decrease the rate of change of x, by decreasing the rate in proportion to the amount of u and x. Let $u(t)$ be the amount of this substance added at time t. For example, the population could be an infestation of an insect, or a harmful microbe in the body. Here we view $x(t)$ as the concentration of a mold and $u(t)$ a fungicide known to kill it. The differential equation representing the mold is given by

$$x'(t) = r(M - x(t)) - u(t)x(t), \quad x(0) = x_0,$$

where $x_0 > 0$ is the given initial population size. Note the term $u(t)x(t)$ pulls down the rate of growth of the mold. The effects of both the mold and fungicide are negative for individuals around them, so we wish to minimize both. Further, while a small amount of either is acceptable, we wish to penalize for amounts too large, so quadratic terms for both will be analyzed. Hence, our problem is as follows

$$\min_u \int_0^T Ax(t)^2 + u(t)^2 \, dt$$

subject to $x'(t) = r(M - x(t)) - u(t)x(t), \quad x(0) = x_0.$

The coefficient A is the weight parameter, balancing the relative importance of the two terms in the objective functional. As we saw in the last lab, one weight term can be divided out, so only the A parameter is needed here. The other parameter in front of the u^2 is taken to be 1. To begin, type *lab2* and press enter. Enter the values

$$\boxed{r = 0.3 \quad M = 10 \quad A = 1 \quad x_0 = 1 \quad T = 5}. \tag{6.1}$$

Do not vary any parameters for now. The control initially increases, then levels off to become constant. The state is also constant here; we say the control and state are in *equilibrium*, meaning both stay at constant values. The control eventually begins decreasing again, going all the way to 0. The state never decreases, with heavy growth at the beginning and end of the interval and constant in the middle. In application, though, we wanted to eliminate the state, or at least decrease it. Note, we entered the value $A = 1$, meaning lowering the level of mold is as important as keeping the levels of fungicide down. This generally would not be the case, however. We are much more interested in removing the mold. Therefore, we should use a higher weight parameter.

Enter the values

$$\boxed{r = 0.3 \quad M = 10 \quad A = 10 \quad x_0 = 1 \quad T = 5}. \tag{6.2}$$

Here, the level of fungicide used is much higher. Notice that the state and control still experience the long period of equilibrium. The control begins at its greatest point, decreasing slightly before becoming constant, then decreasing to 0. As desired, the state decreases from its initial amount to about 0.95 and becomes constant. However, at the end of the interval, when the fungicide use decreases, the level of mold rapidly increases. Seemingly, the best course of action would be to begin another 5-day regimen of a second fungicide on about day 4. For comparison, see Figure 6.1, which shows the optimal state with these values, versus a mold population where no fungicide is used ($u \equiv 0$).

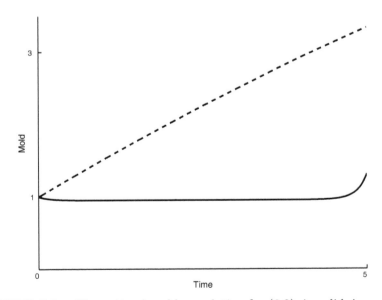

FIGURE 6.1: The optimal mold population for (6.2), in solid, increases at the end of the interval, but is held much lower overall than if no fungicide was used (dashed).

Now try varying two larger values of A. Enter the values (6.2), varying with $A = 15$. The two systems have similar dynamics. A stronger regimen of fungicide is used in the second system; hence, the state is driven lower before becoming constant. Both controls decrease to 0, so that both states experience rapid growth at the end of the period. The second state does not grow as large as the first, though, because it begins at a lower equilibrium point and is still receiving slightly more fungicide.

Enter (6.2), varying with $r = 0.1$. The mold in the second system has slow natural growth. Much less fungicide is used, but the mold in the second system still decreases more than the first. Also, at the end of the interval when both controls decrease, the increase of the first state is a great deal sharper or more rapid than the second. We also mention that the second state is the first we have seen where the amount of mold is less at the end of the interval than at the beginning. Now vary with $r = 0.5$. Here, the state is everywhere increasing, even though a stronger control is used. We are not able to find an acceptable schedule of fungicide strong enough to overcome the natural growth rate of the mold. It is worth noting the two controls begin at the same value.

Examine the carrying capacity M. Enter (6.2), varying with $M = 12$. Clearly, the higher carrying capacity will cause the mold in the second system to naturally increase more rapidly. Much like we saw with the growth rate, the best strategy is to balance this effect with the control and state. Namely, a stronger control is used and a less desirable state is achieved. Note, the controls begin at the same point here as well.

So far, we have only looked at systems where the mold concentration begins at a fraction of carrying capacity. Look at a simulation where x_0 is close to M. Then, try $x_0 = M$. For instance, run

$$\boxed{r = 0.6 \quad M = 5 \quad A = 10 \quad x_0 = 5 \quad T = 5}. \tag{6.3}$$

You will notice a change in the overall behavior of the control. Here, the control begins with an extremely strong dose of fungicide, with $u(0) \approx 5 \times u(1)$. It then quickly returns to the levels we have seen, becoming constant. However, the state, near carrying capacity, will experience little initial natural growth. So, the initial blitz of fungicide is devastating to the population. This allows the large decrease in fungicide use before the equilibrium period, despite the high growth rate.

Now try varying the initial concentration. Enter (6.2), varying with $x_0 = 2$. In this simulation, a much stronger control is used initially in the second system, pushing the second state closer to the first. At approximately $t = 0.75$, the states become identical, as do the controls. Now vary with $x_0 = 3$. Now with $x_0 = 0.5$. We see the same behavior occurs, always in about the same amount of time. So, initial concentration affects only the initial dose of the fungicide regimen. Afterwards, a uniform schedule is used, based on the other parameters. We will see that this phenomena occurs in several of the later labs as well.

Finally, vary the length of the time interval. Enter (6.2), varying with $T = 2.5$. The initial dynamics are identical. The systems differ only after the second system exits the equilibrium state. Notice, the decrease of the control and increase of the state, which occur at the end of the interval, is the same in both systems, only occurring at different times. In fact, if you continue to experiment, you will see altering T only changes the length of the equilibrium period. On that note, if you make T small enough, say $T = 0.5$, you will eliminate the equilibrium period entirely, and the dynamics will be noticeably altered.

Before finishing, we bring up an example of what can go wrong with numerical solutions. MATLAB, due to the amount of information it is able to store, actually has a "largest number." If we enter values which cause x to grow too fast, the state can actually reach this limit. When this happens, MATLAB will simply stop calculating. Enter the values

$$\boxed{r = 1 \quad M = 20 \quad A = 10 \quad x_0 = 1 \quad T = 5}. \tag{6.4}$$

You see the graph goes straight up, then stops at about $t = 1.75$. This behavior also occurs in the next lab.

Chapter 7

Lab 3: Bacteria

Suppose a certain bacteria is grown in a lab, perhaps for medical use. Left alone, the bacteria population will grow exponentially, with growth rate r. A chemical nutrient is known to speed the reproduction process of the bacteria when added. However, use of the chemical by the bacteria creates a second chemical byproduct, which hinders growth. It is also known that the level of hinderance is related to the size of the bacteria population. Namely, the larger the bacteria population is, the smaller the effect this byproduct will have. It is believed this relation is roughly exponential. Therefore, if $x(t)$ is the bacteria concentration at time t, then the growth is given by

$$x'(t) = rx(t) + Au(t)x(t) - Bu(t)^2 e^{-x(t)},$$

where $u(t)$ is the amount of the chemical being added at time t, A is the relative strength of the chemical nutrient increasing growth, and B is the strength of the byproduct. Let $x_0 > 0$ be the given initial concentration. We will consider growth and supplementation over the normalized time interval $[0, 1]$. We wish to maximize x at the end of this interval while simultaneously minimizing the amount of chemical agent used. Thus, our problem can be stated

$$\max_u \ Cx(1) - \int_0^1 u(t)^2 \, dt$$

subject to $\quad x'(t) = rx(t) + Au(t)x(t) - Bu(t)^2 e^{-x(t)}, \ x(0) = x_0,$
$\quad\quad\quad\quad\quad A, B, C \geq 0.$

Before beginning, we make two short notes. First, it is easily shown from the adjoint and transversality conditions that $\lambda(t) > 0$ for all t. Thus, we can get the characterization of the control as usual (see Exercise 7.1). Second, unlike the previous labs, there is a payoff term. Here, $\phi(x) = Cx$ and $\phi' = C$. So, the adjoint is not zero at the end of the interval, but $\lambda(1) = C$. In previous MATLAB codes, we had set the variable λ equal to a vector of zeros, to declare its size. We were also inserting the transversality condition: the adjoint is zero at the final time. Here, we must set $\lambda(1)$ equal to the constant C. You can see in the file *code3.m* this is precisely what is done.

To begin the program, type *lab3* and press enter. Enter the values

$$\boxed{r = 1 \quad A = 1 \quad B = 12 \quad C = 1 \quad x_0 = 1}. \tag{7.1}$$

For now, do not vary any parameters. We see the chemical injection is concentrated at the end of the time interval. This is due to the decreasing effect of the byproduct. As x becomes larger, the e^{-x} term decreases, and the byproduct has less of a hindering effect. Consequently, the level of chemical added starts fairly low and steadily increases, with noticeably higher rates of increase around $t = 0.6$ and $t = 0.8$. As such, the bacteria growth is approximately exponential early in the time interval, but begins to increase more and more rapidly.

Enter (7.1), varying with $x_0 = 0.9$. The two solutions begin very close to each other. However, the chemical use in the first system increases slightly faster than the second, leading to a significant difference by the end of the interval. As the bacteria concentration in the first system becomes larger than the second, the effect of the byproduct becomes less significant, and more chemical can be used. Now try varying with $x_0 = 1.1$. The differences in this simulation are more pronounced than before. As x_0 is increased, more chemical can be used earlier. Now try $x_0 = 1.1$ vs. $x_0 = 1.1495$. This small change almost doubles the final bacteria population. As x_0 inches up, the bacteria population will explode. In fact, entering only $x_0 = 1.16$ will cause the population to grow beyond MATLAB's "largest number," as in the last lab.

Try a small initial population, such as

$$\boxed{r = 1 \quad A = 1 \quad B = 12 \quad C = 1 \quad x_0 = 0.1} \tag{7.2}$$

without varying any parameters. You see very little chemical is used. Due to the small initial count, the population never gets large enough for the byproduct to be as insignificant as before. Now try $x_0 = 0.0001$. Virtually no chemical is used.

Now examine the role of A. Enter (7.1) again, varying with $A = 1.1$. The chemical now aids the growth more. As expected, more chemical is used. However, almost the same amount is used in both systems until about $t = 0.4$. At this point, the byproduct's effect has apparently reached a threshold where the rate of chemical use should be increased more quickly. The positive effect of the chemical is greater in the second system, so more chemical is used there. Similar to what we saw above, as A is increased, the bacteria population will explode. Only $A = 1.4$ is needed to breach MATLAB's limit.

If we decrease A, the chemical will have less positive effect, and less will be used. With $A = 0.4$, a moderate amount is used, while almost none is used when $A = 0.01$. If $A = 0$, then the chemical has no positive effect at all. It is broken down by the bacteria, with no benefit, to create a harmful byproduct. Not surprisingly, no chemical is used when $A = 0$, regardless of the other parameters.

Now enter (7.1), varying with $B = 20$. As expected, with more harmful byproduct, less chemical is used. It is worth noting that when A was adjusted, the use in chemical would stay almost the same for much of the beginning of the time interval, before increasing rapidly. Here, the chemical uses in the two systems begin apart, and the difference between them steadily increases. Like before, if we continue to increase B, less and less chemical will be used.

Of interest is the behavior of the control as B decreases. Try

$$\boxed{r = 1 \quad A = 1 \quad B = 0.1 \quad C = 1 \quad x_0 = 0.1} . \qquad (7.3)$$

The control is now concave, where as in most of the other simulations we have done it has been convex. For smaller B, with appropriate x_0 so that MATLAB can handle the numbers, we see the control has less variation. The extreme of this occurs when $B = 0$. Enter

$$\boxed{r = 1 \quad A = 1 \quad B = 0 \quad C = 1 \quad x_0 = 0.01} . \qquad (7.4)$$

Here, the control appears constant. As there is no negative byproduct, the optimal control is one of almost constant chemical injection. (Note: The control is most likely not exactly constant. The overall change is simply so small that MATLAB's graphing tools cannot display it. You may have been given an error message to this effect.)

Examine the role of the growth rate. Enter (7.1), varying with $r = 1.1$. In the second system, the bacteria has a higher natural growth rate. Therefore, the byproduct becomes less harmful more quickly, and more chemical can be used. Notice that the optimal control, and thus state, are virtually the same until about $t = 0.4$. Now compare $r = 1.1$ to $r = 1.2$. As r is increased, the control becomes more varied and reaches higher maximum levels. Conversely, as r is decreased, the control has less variation. Try $r = 0.8$. The control here experiences far less rapid growth at the end of the interval. Now try $r = 0.1$. The control is almost linear.

Finally, experiment with the weight parameter C. So far, we have used $C = 1$, meaning maximizing the final bacterial concentration and minimizing total chemical usage are of equal importance. Suppose the chemical is cheap and plentiful, and we are not very concerned with how much we use. Compare $C = 1$ vs. $C = 5$, using (7.1). More chemical is used to drive the bacterial concentration higher. The two systems differ, though, only after about $t = 0.7$. On the other hand, suppose the chemical is very expensive, and we are only willing to use a little to adjust the bacterial growth. Try $C = 1$ against $C = 0.2$. As expected, less chemical agent is used, although the effect of the weaker chemical schedule does not become apparent immediately.

Exercise 7.1 Calculate the necessary conditions for this lab problem. Using the adjoint equation and transversality condition, show $\lambda(t) > 0$ for all t, so that the characterization of the optimal control

$$u^* = \frac{Ax\lambda}{2(1 + B\lambda e^{-x})}$$

is well-defined.

Chapter 8

Bounded Controls

Many problems require bounds on the control to achieve a realistic solution. Suppose, for instance, that our control is the amount of a chemical used in a system. Then, clearly we require this amount to be nonnegative, i.e., $u \geq 0$. Often, the control must also be bounded above. Perhaps there are physical limitations on the amount of chemicals or environmental regulations which prohibit a certain level of use. We could also have a problem where the control is the percentage of some strength or use. Then $0 \leq u \leq 1$ would be our bounds.

Recall that in Labs 2 and 3, the controls were a fungicide and the concentration of a chemical nutrient, respectively. Clearly, both of these quantities must remain non-negative. We did not, however, enforce this with bounds in the problem, as the resulting optimal controls met this requirement without restriction. However, this is not always the case. Consider the following fish harvesting example.

Example 8.1 (from [40])

We wish to maximize net profit of harvested fish,

$$\max_u \int_0^T \left(p_1 u(t) x(t) - p_2 u(t)^2 x(t)^2 - c u(t) \right) dt$$
$$\text{subject to} \quad x'(t) = K x(t)(M - x(t)) - u(t) x(t), \ x(0) = x_0,$$

where $x(t)$ is the population concentration of the fish, $u(t)$ is the level of harvesting, p_1, p_2, and c are the terms representing revenue from sale of fish, diminishing returns when there is a large amount of fish to sell, and cost of fishing. The amount of fish harvested at time t is $u(t)x(t)$ and using the price p_1, the first revenue term is $p_1 u(t) x(t)$. The variables M and K are the carrying capacity and growth rate of the fish population, respectively.

Calculating the necessary conditions from the Hamiltonian, we obtain

$$H = p_1 ux - p_2(ux)^2 - cu + \lambda(Kx(M-x) - ux),$$
$$\lambda' = -\frac{\partial H}{\partial x} = -[p_1 u - 2p_2 u^2 x + \lambda(KM - 2Kx - u)], \ \lambda(T) = 0,$$
$$0 = \frac{\partial H}{\partial u} = p_1 x - 2p_2 ux^2 - c - \lambda x \Rightarrow$$
$$u^* = \frac{-\lambda x^* + p_1 x^* - c}{2p_2 (x^*)^2}.$$

One can see the need to solve for u^*, x^*, λ numerically. For certain, plausible values of the constants, the optimal control will be negative during the time interval. For example, see the optimal controls in Figure 8.1. Clearly, this is physically impossible. The optimal control in this case should be the control u^* that maximizes the objective functional, chosen only from controls u such that $u \geq 0$. Thus for this example, a lower bound constraint on the controls would be essential. For many applications, upper and lower bounds on the controls would be reasonable.

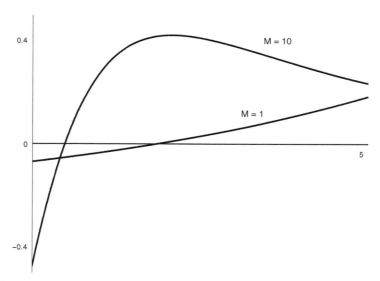

FIGURE 8.1: Two optimal controls for Example 8.1, with $p_1 = 2$, $p_2 = 1$, $c = 1$, $K = 0.25$, $x_0 = 0.5$, and $T = 5$. One control is for $M = 1$ and the other for $M = 10$. Both optimal controls are negative for part of the interval.

8.1 Necessary Conditions

In order to solve problems with bounds on the control, we must develop alternate necessary conditions. Consider the problem

$$\max_u \int_{t_0}^{t_1} f(t, x(t), u(t))\, dt + \phi(x(t_1))$$

subject to $\quad x'(t) = g(t, x(t), u(t)),\ x(t_0) = x_0,$
$$a \leq u(t) \leq b,$$

where a, b are fixed, real constants and $a < b$. Let $J(u)$ be the value of the objective functional at control u, where $x = x(u)$ is the associated state, namely,

$$J(u) = \int_{t_0}^{t_1} f(t, x(t), u(t))\, dt + \phi(x(t_1)).$$

Let u^*, x^* be an optimal pair. Let $h(t)$ be a piecewise continuous function where there exists a positive constant ϵ_0, such that for all $\epsilon \in (0, \epsilon_0]$, $u^\epsilon(t) = u^*(t) + \epsilon h(t)$ is admissible, i.e.,

$$a \leq u^\epsilon(t) \leq b \quad \text{for all} \quad t.$$

Due to bounds on the controls, the derivative of the objective functional may not be zero at the optimal control, since u^* may be at the bounds (endpoints of its range) at some points in time; we may only know the sign of this derivative. To calculate this sign, we also restrict the sign of the ϵ parameter. Let $x^\epsilon(t)$ be the corresponding state variable for each $\epsilon \in (0, \epsilon_0]$. Precisely as was done in Chapter 1, introduce a piecewise differentiable adjoint variable $\lambda(t)$ and apply the Fundamental Theorem of Calculus to write $J(u^\epsilon)$ as

$$J(u^\epsilon) = \int_{t_0}^{t_1} \left[f(t, x^\epsilon, u^\epsilon) + \lambda(t) g(t, x^\epsilon, u^\epsilon) + x^\epsilon(t) \lambda'(t) \right] dt \qquad (8.1)$$
$$- \lambda(t_0) x_0 + \lambda(t_1) x^\epsilon(t_1) + \phi(x(t_1)).$$

As the maximum of $J(u)$ with respect to u occurs at u^*,

$$0 \geq \frac{d}{d\epsilon} J(u^\epsilon) \bigg|_{\epsilon=0} = \lim_{\epsilon \to 0^+} \frac{J(u^\epsilon) - J(u^*)}{\epsilon}. \qquad (8.2)$$

Note, the constant ϵ was chosen to be positive, so the limit can only be taken from one side. The numerator is clearly non-positive, as u^* is maximal. This gives the inequality shown, instead of equality as in Chapter 1. However, this is all we will need. As we did before, choose the adjoint variable so that

$$\lambda'(t) = -\big[f_x(t, x^*, u^*) + \lambda(t)g_x(t, x^*, u^*)\big], \quad \lambda(t_1) = \phi'(x^*(t_1)).$$
Then (8.1), (8.2) reduce to

$$0 \geq \int_{t_0}^{t_1} (f_u + \lambda g_u) h\, dt, \tag{8.3}$$

and this inequality holds for all h as described above.

Let s be a point of continuity of u^* with $a \leq u^*(s) < b$. Suppose $f_u + \lambda g_u > 0$ at s. As u^* is continuous at s, so is $f_u + \lambda g_u$. Thus, there is a small interval I, containing s, on which $f_u + \lambda g_u$ is strictly positive and $u^* < b$. Let

$$M = \max\{u^*(t) : t \in I\} < b.$$

Define a particular h by

$$h(t) = \begin{cases} b - M & \text{if } t \in I, \\ 0 & \text{if } t \notin I. \end{cases}$$

Note, $h > 0$ on I. Further, it is easily seen that $a \leq u^* + \epsilon h \leq b$ for all $\epsilon \in [0, 1]$. But,

$$\int_{t_0}^{t_1} (f_u + \lambda g_u) h\, dt = \int_I (f_u + \lambda g_u) h\, dt > 0$$

by choice of I and h. This contradicts (8.3), and implies $f_u + \lambda g_u \leq 0$ at s.

Similarly, let s be a point of continuity of u^* with $a < u^*(s) \leq b$. Suppose $f_u + \lambda g_u < 0$ at s. As before, there is a small interval I, containing s, on which $f_u + \lambda g_u$ is strictly negative and $u^* > a$. Let $m = \min\{u^*(t) : t \in I\}$, and define a variation function by $h = a - m$ on I and 0 off I. Then, $a \leq u^* + \epsilon h \leq b$ for all $\epsilon \in [0, 1]$. But,

$$\int_{t_0}^{t_1} (f_u + \lambda g_u) h\, dt = \int_I (f_u + \lambda g_u) h\, dt > 0,$$

which contradicts (8.3). So, $f_u + \lambda g_u \geq 0$ at s. Further, this holds for all points of continuity s. In summary,

$$\begin{aligned} u^*(t) = a & \quad \text{implies} \quad f_u + \lambda g_u \leq 0 \text{ at } t, \\ a < u^*(t) < b & \quad \text{implies} \quad f_u + \lambda g_u = 0 \text{ at } t, \\ u^*(t) = b & \quad \text{implies} \quad f_u + \lambda g_u \geq 0 \text{ at } t. \end{aligned} \tag{8.4}$$

The conditions (8.4) are equivalent to

$$\begin{aligned} f_u + \lambda g_u < 0 \text{ at } t & \quad \text{implies} \quad u^*(t) = a, \\ f_u + \lambda g_u = 0 \text{ at } t & \quad \text{implies} \quad a \leq u^*(t) \leq b, \\ f_u + \lambda g_u > 0 \text{ at } t & \quad \text{implies} \quad u^*(t) = b. \end{aligned} \tag{8.5}$$

Bounded Controls

This holds for all points of continuity t of u^*. As they are irrelevant to the objective functional and the state equation, we neglect the remaining points. These new necessary conditions can be compiled as before. Forming the Hamiltonian

$$H(t, x, u, \lambda) = f(t, x, u) + \lambda(t)g(t, x, u),$$

the necessary conditions for x^* and λ are unchanged, namely

$$x'(t) = \frac{\partial H}{\partial \lambda}, \quad x(t_0) = x_0,$$

$$\lambda'(t) = -\frac{\partial H}{\partial x}, \quad \lambda(t_1) = \phi'(x(t_1)).$$

It follows from the derivation above

$$\begin{cases} u^* = a & \text{if } \frac{\partial H}{\partial u} < 0 \\ a \leq u^* \leq b & \text{if } \frac{\partial H}{\partial u} = 0 \\ u^* = b & \text{if } \frac{\partial H}{\partial u} > 0. \end{cases} \tag{8.6}$$

A version of Pontryagin's Maximum Principle is also true here. It is essentially the same as that of Theorem 1.2, except the maximization is over all *admissible* controls, i.e., all controls which adhere to the bounds. In particular, the optimal control u^* maximizes H pointwise with respect to $a \leq u \leq b$. If we have a minimization problem, then u^* is instead chosen to minimize H pointwise. This has the effect of reversing $<$ and $>$ in the first and third lines of (8.6). See Exercise 8.1.

Note that the bounds on the control had no effect on the transversality condition. In developing the above necessary conditions, we dealt only with the case of the state being fixed at the initial time and free at the terminal time. However, the other cases are handled just as before. For instance, if $x(t_0)$ and $x(t_1)$ are both fixed, then the adjoint variable will have no boundary conditions. We now present the following examples with constraints on the controls.

Example 8.2 (from [100])

$$\max_u \int_0^2 \left[2x(t) - 3u(t) - u(t)^2 \right] dt$$

$$\text{subject to} \quad x'(t) = x(t) + u(t), \ x(0) = 5,$$
$$0 \leq u(t) \leq 2.$$

Form the Hamiltonian

$$H = 2x - 3u - u^2 + x\lambda + u\lambda.$$

Then, the adjoint calculation yields:

$$\lambda'(t) = -\frac{\partial H}{\partial x} = -2 - \lambda \Rightarrow \lambda = -2 + c_1 e^{-t}$$
$$\lambda(2) = 0 \Rightarrow c_1 = 2e^2 \Rightarrow \lambda = 2e^{2-t} - 2.$$

Now that we have found the adjoint value, we turn our attention to u^*, which requires considering the sign of $\frac{\partial H}{\partial u}$:

$$\frac{\partial H}{\partial u} = -3 - 2u + \lambda,$$

$$0 > \frac{\partial H}{\partial u} \text{ at } t \Rightarrow u(t) = 0 \Rightarrow 0 > -3 + \lambda = -3 + (2e^{2-t} - 2)$$
$$\Rightarrow t > 2 - \ln(5/2),$$

$$0 < \frac{\partial H}{\partial u} \text{ at } t \Rightarrow u(t) = 2 \Rightarrow 0 < -3 - 2(2) + \lambda = -7 + (2e^{2-t} - 2)$$
$$\Rightarrow t < 2 - \ln(9/2),$$

$$0 = \frac{\partial H}{\partial u} \text{ at } t \Rightarrow u(t) = \frac{1}{2}(\lambda - 3) \Rightarrow 0 \leq \frac{1}{2}(\lambda - 3) \leq 2$$
$$\Rightarrow 2 - \ln(9/2) \leq t \leq 2 - \ln(5/2).$$

Hence, the optimal control is

$$u^*(t) = \begin{cases} 2 & \text{when } 0 \leq t < 2 - \ln(\tfrac{9}{2}), \\ e^{2-t} - \tfrac{5}{2} & \text{when } 2 - \ln(\tfrac{9}{2}) \leq t \leq 2 - \ln(\tfrac{5}{2}), \\ 0 & \text{when } 2 - \ln(\tfrac{5}{2}) < t \leq 2. \end{cases}$$

To find the optimal state, insert the values for u^* into the differential equation for x, and solve the three cases. We find the optimal state to be

$$x^*(t) = \begin{cases} k_1 e^t - 2 & \text{when } 0 \leq t < 2 - \ln(\tfrac{9}{2}), \\ k_2 e^t - \tfrac{1}{2} e^{2-t} + \tfrac{5}{2} & \text{when } 2 - \ln(\tfrac{9}{2}) \leq t \leq 2 - \ln(\tfrac{5}{2}), \\ k_3 e^t & \text{when } 2 - \ln(\tfrac{5}{2}) < t \leq 2, \end{cases}$$

where k_1, k_2, and k_3 are constants. Using $x(0) = 5$, it follows $k_1 = 7$. Recall, the state must be continuous. So, requiring x^* to agree at $t = 2 - \ln(\tfrac{9}{2})$ and $t = 2 - \ln(\tfrac{5}{2})$, we find values for k_2 and k_3, so that

$$x^*(t) = \begin{cases} 7e^t - 2 & \text{when } 0 \leq t \leq 2 - \ln(\tfrac{9}{2}), \\ (7 - \tfrac{81}{8} e^{-2}) e^t - \tfrac{1}{2} e^{2-t} + \tfrac{5}{2} & \text{when } 2 - \ln(\tfrac{9}{2}) \leq t \leq 2 - \ln(\tfrac{5}{2}), \\ (7 - 7e^{-2}) e^t & \text{when } 2 - \ln(\tfrac{5}{2}) \leq t \leq 2. \end{cases}$$

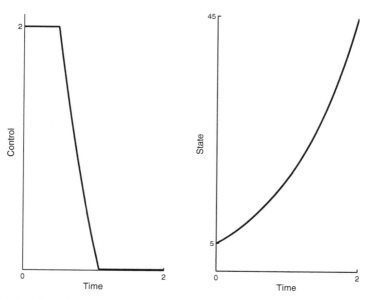

FIGURE 8.2: Optimal control and state for Example 8.2.

Figure 8.2 displays the optimal control and state for this example.

Example 8.3 This example deals with a one-sided control constraint.

$$\max_{u} \; x(4) - \int_0^4 u(t)^2 \, dt$$
$$\text{subject to} \quad x'(t) = x(t) + u(t), \; x(0) = 0,$$
$$u(t) \leq 5.$$

The Hamiltonian in this problem is

$$H = -u^2 + \lambda x + \lambda u.$$

Recall, the payoff term $\phi = x(4)$ is not included in the Hamiltonian, but instead incorporated into the transversality condition. Specifically, since $\phi(x) = x$ and $\phi' = 1$, we have

$$\lambda'(t) = -\frac{\partial H}{\partial x} = -\lambda, \quad \lambda(4) = \phi'(x^*(4)) = 1.$$

This gives

$$\lambda(t) = e^{4-t}.$$

If you refer back to equation (8.5), you will see $\frac{\partial H}{\partial u} < 0$ implies u^* is at the lower bound. However, we have no lower bound in this problem. The control u can range over all values less than or equal to 5. So, $\frac{\partial H}{\partial u} < 0$ cannot occur. To find a representation of u^*, we need only consider the other two cases:

$$\frac{\partial H}{\partial u} = \lambda - 2u,$$

$\frac{\partial H}{\partial u} > 0 \Rightarrow u^*(t) = 5 \Rightarrow \lambda - 10 > 0 \Rightarrow e^{4-t} > 10 \Rightarrow t < 4 - \ln(10),$

$\frac{\partial H}{\partial u} = 0 \Rightarrow u^*(t) \leq 5 \Rightarrow e^{4-t} = \lambda = 2u \leq 10 \Rightarrow t \geq 4 - \ln(10).$

Hence, the above two cases give

$$u^* = \begin{cases} 5 & \text{when } 0 \leq t < 4 - \ln(10), \\ \frac{1}{2}e^{4-t} & \text{when } 4 - \ln(10) \leq t \leq 4. \end{cases}$$

To finish the example, we simply solve the state equation to find x^*:

$$x'(t) = x + 5, \; x(0) = 0 \Rightarrow x(t) = 5e^t - 5 \text{ on } [0, 4 - \ln(10)],$$

$$x'(t) = x + \frac{1}{2}e^{4-t} \Rightarrow x(t) = -\frac{1}{4}e^{4-t} + ke^t \text{ on } [4 - \ln(10), 4],$$

for some constant k. We require that x^* be continuous, so these two expressions must agree at $t = 4 - \ln(10)$. This gives $k = 5 - 25e^{-4}$. Hence,

$$x^* = \begin{cases} 5e^t - 5 & \text{when } 0 \leq t \leq 4 - \ln(10) \\ -\frac{1}{4}e^{4-t} + (5 - 25e^{-4})e^t & \text{when } 4 - \ln(10) \leq t \leq 4. \end{cases}$$

Example 8.4 The following example illustrates an important concept. Namely, an optimal control cannot be found by ignoring the bounds while solving the necessary conditions, then truncating the result. Refer back to Example 3.5 and consider the same problem with bounds on the control.

$$\min_u \int_0^4 u(t)^2 + x(t) \, dt$$

subject to $x'(t) = u(t), \; x(0) = 0, \; x(4) = 1,$
$u(t) \geq 0.$

Using the Hamiltonian

$$H = u^2 + x + \lambda u,$$

we find the adjoint function
$$\lambda'(t) = -\frac{\partial H}{\partial x} = -1 \quad \Rightarrow \quad \lambda(t) = k - t,$$
for some constant k. To find u^*, minimize H pointwise:
$$\frac{\partial H}{\partial u} = 2u + \lambda,$$
$$\frac{\partial H}{\partial u} > 0 \Rightarrow u^*(t) = 0 \Rightarrow 0 < \lambda = k - t \Rightarrow t < k,$$
$$\frac{\partial H}{\partial u} = 0 \Rightarrow 0 \le u^* = -\frac{\lambda}{2} = \frac{t-k}{2} \Rightarrow t \ge k.$$
So, we have the representation
$$u^*(t) = \begin{cases} 0 & \text{when} \quad 0 \le t < k, \\ \frac{t-k}{2} & \text{when} \quad k \le t \le 4. \end{cases} \tag{8.7}$$
So, we need only find the value for k.

Case 1: $k \le 0$. Here, $0 \le t < k$ is impossible. So, by (8.7), $x'(t) = u = \frac{t-k}{2}$ on $[0,4]$. Thus, using $x(0) = 0$, the state differential equation yields $x(t) = \frac{t^2}{4} - \frac{kt}{2}$. However, $1 = x(4) = 4 - 2k$, which implies $k = \frac{3}{2} > 0$. Contradiction.

Case 2: $k \ge 4$. Now $k \le t \le 4$ is impossible, so that $u^* \equiv 0$. Then, $x'(t) = u = 0$ on $[0,4]$. Thus, $x(0) = 0$ gives $x(t) = 0$ for all t. However, $1 = x(4) = 0$. Contradiction.

Case 3: $0 < k < 4$. By (8.7), $x'(t) = u = 0$ on $[0,k)$. The initial condition $x(0) = 0$ implies $x(t) = 0$ on $[0,k)$. On $[k,4]$, $x'(t) = \frac{t-k}{2}$. This gives
$$x(t) = \frac{t^2}{4} - \frac{k}{2}t + c,$$
for some constant c. From the continuity of x, $c = \frac{k^2}{4}$, so that $x(t) = \frac{(t-k)^2}{4}$ on $[k,4]$. Finally, $1 = x(4) = \frac{(4-k)^2}{4}$, which implies $k = 2$ or 6. We assumed $k < 4$, so $k = 2$.

Hence the optimal solutions are
$$u^*(t) = \begin{cases} 0 & \text{when} \quad 0 \le t < 2, \\ \frac{t-2}{2} & \text{when} \quad 2 \le t \le 4, \end{cases} \quad \text{and} \quad x^*(t) = \begin{cases} 0 & \text{when} \quad 0 \le t \le 2, \\ \frac{(t-2)^2}{4} & \text{when} \quad 2 \le t \le 4. \end{cases}$$

Suppose we truncated the optimal control u^* from Example 3.5, to form a new control $\hat{u}$, where $\hat{u} \ge 0$. Then,
$$\hat{u}(t) = \begin{cases} 0 & \text{when} \quad 0 \le t \le \frac{3}{2}, \\ \frac{2t-3}{4} & \text{when} \quad \frac{3}{2} \le t \le 4. \end{cases}$$

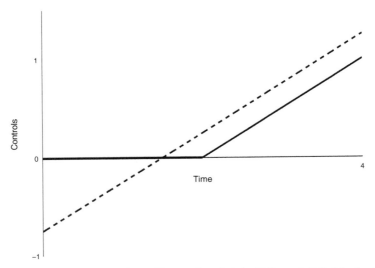

FIGURE 8.3: Controls from Examples 8.4 (solid) and 3.5 (dashed). We see the first is not the truncation of the second.

As you can see, $\hat{u}$ and u^* from Example 8.4 are not the same. The bounds must be taken into account while solving the necessary conditions, not just used to truncate the solution. Figure 8.3 shows the optimal control from this example versus that from Example 3.5. Lab 4 deals with the idea of truncation further.

Although it is beyond the scope of this book, it is worth mentioning that more complicated constraints can be placed on the control. In general, one can consider control constraints of the form

$$h_j(t, u(t)) \geq 0, \quad j = 1, ..., p,$$

where each h_j is continuously differentiable in both variables. This situation might arise if one wants to constrain the range values of the controls to be in a certain set, rather than just lie between two bounds. This can be done using the h_j functions. In fact, constraints which also involve the state can be added to the problem. For example, one might want to restrict the level of cancer cells by an upper constraint, $x(t) \leq M$. For more information on this, refer to [100, 130, 169].

8.2 Numerical Solutions

In order to numerically solve a problem with the addition of bounds, only a slight modification of the forward-backward sweep routine is needed. As before, an initial guess of the control is made. The state equation and adjoint equation are not directly affected by the bounds, so the state and adjoint are solved forward and backward in time, respectively, as before. Only the characterization of the control is changed. It must be altered to reflect the bounds. Before, u^* was identically equal to some expression in t, x, and/or λ. Now u can be equal to that expression or set at one of the bounds; this is a type of truncation. There is a concise way of entering these cases into the code, which we will examine via an example.

Consider the problem we first used to motivate the routine in Chapter 4, this time with bounds:

$$\max_u \int_0^1 Ax(t) - u(t)^2 \, dt$$

subject to $\quad x'(t) = -\frac{1}{2}x(t)^2 + Cu(t), \; x(0) = x_0 > -2,$

$$M_1 \leq u(t) \leq M_2.$$

Notice one of the weight parameters B has been removed from the problem. Recall, we found that the second weight parameter was superfluous. Here, the Hamiltonian is

$$H = Ax - u^2 - \frac{1}{2}x^2\lambda + Cu\lambda.$$

The adjoint and transversality conditions are

$$\lambda'(t) = -\frac{\partial H}{\partial x} = -A + x\lambda, \quad \lambda(1) = 0.$$

We find a characterization of u^* by considering three cases:

$$\frac{\partial H}{\partial u} = -2u + C\lambda,$$
$$\frac{\partial H}{\partial u} < 0 \Rightarrow u(t) = M_1 \Rightarrow M_1 > \frac{C\lambda}{2},$$
$$\frac{\partial H}{\partial u} = 0 \Rightarrow u(t) = \frac{C\lambda}{2} \Rightarrow M_1 \leq \frac{C\lambda}{2} \leq M_2,$$
$$\frac{\partial H}{\partial u} > 0 \Rightarrow u(t) = M_2 \Rightarrow M_2 < \frac{C\lambda}{2}.$$

Notice the recurrence of the expression $\frac{C\lambda}{2}$. When the control is at the upper bound, this expression is strictly greater than M_2. Similarly, when the control is at the lower bound, the expression is strictly less than M_1. Thus, a compact way of writing the optimal control is

$$u^*(t) = \min\left(M_2, \max\left(M_1, \frac{C\lambda(t)}{2}\right)\right).$$

This is merely a truncation of the expression $\frac{C\lambda}{2}$ with the upper and lower bounds. Note, however, that this truncation occurs within the problem. The control is equal to a truncated expression in $\frac{C\lambda}{2}$, but the state and adjoint are directly affected by u. This is very different from solving the problem without bounds and truncating the resulting solution, as we have seen.

To alter the forward-backward code to reflect the bounds, we only have to change the characterization of u. Recall that in the routine with no bounds, *code1.m*, the characterization was given by the following.

```
                              code1.m
41    u1 = C*lambda/(2*B);
42    u = 0.5*(u1 + oldu);
```

We simply change to the truncated expression found above (and set $B = 1$, as it was divided out).

```
                              code4.m
41    u1 = min(M2, max(M1, C*lambda/2));
42    u = 0.5*(u1 + oldu);
```

The *lambda* expression in line 38 above is actually a vector. However, the *min* and *max* functions in MATLAB operate term-by-term. In other words, they compare each individual term of the vector, finding the maximum and minimum, thereby creating a new vector. The convex combination is still used in line 39 to speed convergence. For this reason, it can be advantageous to make the initial guess of the control lie within the bounds. However, we continue to use $u \equiv 0$ as the initial guess in all codes. Lab 4 uses this altered code to allow further study of this problem.

Again, it is worth pointing out that there are many different kinds of control updates. Like in Chapter 4, we present a more sophisticated convex combination. We illustrate using the above example. Let $0 < c < 1$ and k the number of iterates. Consider the control update

```
u1 = min(M2, max(M1, C*lambda/2));
for i=1:N+1
    if(u1(i) > oldu(i))
        u(i) = M2*(1 - c^k) + oldu(i)*c^k;
    else
        u(i) = M1*(1 - c^k) + oldu(i)*c^k;
    end
end
```

Here, we use the bounds to dampen the previous iterate, based on the range of the new approximation. Like in Chapter 4, this new method tends to be less accurate than the method we developed earlier, but often works when other methods fail. We will continue to use the simple averaging approach.

8.3 Exercises

Exercise 8.1 Find the necessary conditions for the following optimal control problem:

$$\min_u \int_{t_0}^{t_1} f(t, x(t), u(t))\, dt + \phi(x(t_1))$$

$$\text{subject to} \quad x'(t) = g(t, x(t), u(t)),\ x(0) = x_0,$$
$$a \leq u(t) \leq b.$$

This can be done by emulating the proof given at the beginning of this chapter, or by converting this into a maximization problem by negating the integral.

Exercise 8.2 Solve

$$\max_u \int_0^1 x(t) - u(t)\, dt$$

$$\text{subject to} \quad x'(t) = 2u(t)(1 - u(t)),\ x(0) = 0,$$
$$0 \leq u(t) \leq 1.$$

Exercise 8.3 Solve

$$\max_u \int_0^1 x(t)u(t)\, dt$$

$$\text{subject to} \quad x'(t) = u(t)^2,\ x(1) = -1,$$
$$-1 \leq u(t) \leq 0.$$

Exercise 8.4 (from [100]) The following is a generalization of Example 8.4. Let $c, B, T > 0$ be constants and suppose $B < cT^2/4$. Solve

$$\min_u \int_0^T u(t)^2 + cx(t)\, dt$$

subject to $x'(t) = u(t),\ x(0) = 0,\ x(T) = B,$
$u(t) \geq 0.$

Exercise 8.5 Solve the following problem. There are three different solutions, depending on the values of a and b.

$$\min_u \frac{1}{2}\int_0^1 u(t)^2\, dt + \frac{1}{2}x(1)^2$$

subject to $x'(t) = u(t),\ x(0) = 1,$
$a \leq u(t) \leq b.$

Exercise 8.6 Reconsider Exercise 3.2 with the values $d = S = T = 1$ and with control constraints $0.6 \leq u(t) \leq 0.9$. Namely, solve

$$\max_u \int_0^1 (x(t) - \frac{1}{2}u(t)^2)\, dt + x(1)$$

subject to $x'(t) = u(t) - x(t),\ x(0) = x_0 > 0,$
$0.6 \leq u(t) \leq 0.9.$

Chapter 9

Lab 4: Bounded Case

In this lab, we reexamine the first lab, this time imposing bounds on the control. Also notice that the weight parameter B has been removed from the problem, and only one weight parameter is used, as discussed in Lab 1.

$$\max_u \int_0^1 Ax(t) - u(t)^2 \, dt$$

subject to $x'(t) = -\dfrac{1}{2}x(t)^2 + Cu(t),\ x(0) = x_0 > -2,$

$M_1 \le u(t) \le M_2,\ A \ge 0.$

Open MATLAB and begin *lab4*. In Lab 1, we first examined the optimal control for the parameters values $A = x_0 = 1$ and $C = 4$ (with $B = 1$). There, the optimal control lies between 0 and 2 (it appears the control is bounded by 1, but in fact it has a maximum value slightly above 1). You may want to run this simulation in *lab1* again to refresh your memory. Now, running *lab4*, enter the values

$$\boxed{A = 1 \quad C = 4 \quad x_0 = 1 \quad M_1 = -1 \quad M_2 = 2}. \tag{9.1}$$

Now try

$$\boxed{A = 1 \quad C = 4 \quad x_0 = 1 \quad M_1 = 0 \quad M_2 = 1.5}. \tag{9.2}$$

In both cases, the optimal control is unchanged from Lab 1. If the problem has a solution without bounds, and bounds which contain that solution are added, then the solution will be unchanged, as expected.

Enter

$$\boxed{A = 1 \quad C = 4 \quad x_0 = 1 \quad M_1 = 0 \quad M_2 = 2} \tag{9.3}$$

varying with $M_2 = 0.5$. Clearly, this set of bounds will affect the original solution. The bounded control is very similar to the first control if it were truncated at 0.5. However, if you look closely enough, you will see the second control remains at its upper bound for a short time after the first control passes the bound. You may need to use the zoom tool at the top of the figure window, or expand the figure to full screen, in order to see this. Figure 9.1

also gives a closer view. In addition, the adjoints from the two problems are different, particularly near the beginning of the interval. The effect of the different controls is seen in the states, as the first control, which is not inhibited by the bound, increases the state more.

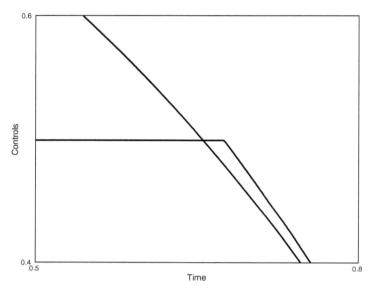

FIGURE 9.1: Optimal control plots for parameters (9.3) together with varying M_2 to be 0.5. We see one control is not a truncation of the other.

For another example, try the values

$$\boxed{A = 20 \quad C = 2 \quad x_0 = 0 \quad M_1 = 0 \quad M_2 = 6}, \tag{9.4}$$

and then vary with $M_2 = 3$. Again, the resulting control is similar to a truncation of the original, but not exactly. The difference should be easier to see than in the last simulation.

Now run a similar experiment on the lower bound. Enter (9.4), this time varying with $M_1 = 5$. Here, the two controls have less similarity than in the previous experiments. As the second control is forced to stay at the relatively high upper bound for the majority of the interval, it begins slightly lower than the first control to compensate.

This time, we will begin with stringent upper and lower bounds. Enter the values

$$\boxed{A = 20 \quad C = 2 \quad x_0 = 0 \quad M_1 = 1 \quad M_2 = 5}. \tag{9.5}$$

Lab 4: Bounded Case

The optimal control will now be affected by both bounds. Vary with $M_1 = 4$. Here, not only does the second control reach its lower bound before the first control, but it also decreases from the mutual upper bound first. Interestingly, if we instead vary with $M_2 = 2$, we find that the second control remains at its upper bound after the first control passes it, as before, but the two controls reach their mutual lower bound at much closer times than in the previous simulation.

Enter the values

$$\boxed{A = 1 \quad C = 4 \quad x_0 = 1 \quad M_1 = 2 \quad M_2 = 3}. \tag{9.6}$$

The original control lies entirely outside these bounds, and the resulting control lies entirely at the lower bound. Now try $M_1 = -2$ and $M_2 = -1$. Here, the control is identically the upper bound. As a special case, enter $M_1 = M_2 = 0$. Of course, the optimal control is $u^* \equiv 0$, as this is the only solution which satisfies the bounds.

Finally, even with the addition of the bounds, the parameters A, C, and x_0 have the same effect as before. For example, enter

$$\boxed{A = 1 \quad C = 4 \quad x_0 = 1 \quad M_1 = 0.25 \quad M_2 = 0.75} \tag{9.7}$$

varying with $A = 3$. The second system, with more emphasis placed on maximizing x, uses a greater control, where possible, in order to decrease the state more. Vary C and x_0 to see they also affect the solution as before.

Chapter 10

Lab 5: Cancer

Optimal control techniques are of great use in developing optimal strategies for chemotherapy [173]. Specifically, a treatment regimen, cast as the control, which will minimize the tumor density and drug side-effects over a given time interval, can be found. This technique was employed by Fister and Panetta in [61]. There, the tumor is assumed to have Gompertzian growth. Several models of chemotherapeutic kill-cell (killing of tumor cells) exist. Three different models are treated in [61]. Here, we examine only one, namely, Skipper's log-kill hypothesis, which states cell-kill due to chemotherapeutic drugs is proportional to tumor population. Thus, if $N(t)$ is the normalized density of the tumor at time t, we have the model

$$N'(t) = rN(t)\ln\left(\frac{1}{N(t)}\right) - u(t)\delta N(t),$$

where r is the growth rate of the tumor, δ is the magnitude of the dose, and $u(t)$ describes the pharmacokinetics of the drug, i.e., $u(t) = 0$ implies no drug effect and $u(t) > 0$ is the strength of the drug effect. The initial condition is taken to be $N(0) = N_0$, where $0 < N_0 < 1$, as the tumor cells have been normalized. The objective functional used is quadratic, where the cost of the control, representing possible side-effects, and the tumor density N are minimized over a time interval. Finally, we require $u(t) \geq 0$ for all t. So, our problem is

$$\min_u \int_0^T aN(t)^2 + u(t)^2\,dt$$

subject to $N'(t) = rN(t)\ln\left(\frac{1}{N(t)}\right) - u(t)\delta N(t),\ N(0) = N_0,$

$u(t) \geq 0.$

Here, a is a positive weight parameter. Enter MATLAB and begin *lab5*. First, try the values

$$\boxed{r = 0.3 \quad a = 3 \quad \delta = 0.45 \quad N_0 = 0.975 \quad T = 20}. \tag{10.1}$$

Notice that the optimal treatment strategy is one of high drug strength early followed by a slow reduction to no drug treatment on day 20. This is consistent with most medical practices today. However, we see the lowered drug strength allows for a slight increase in tumor density after day 12. See Figure 10.1 to compare this treatment to no treatment at all.

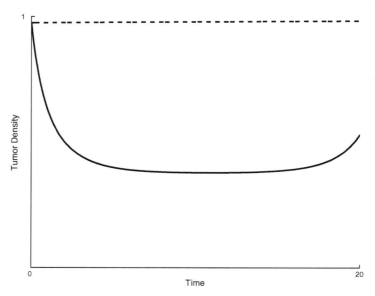

FIGURE 10.1: Tumor density with optimal treatment from (10.1), in solid, compared density with no treatment ($u \equiv 0$) in dashed.

Now, using (10.1), varying a. Use a much higher second a value, say $a = 10$. Notice we are able to push the tumor density to a much lower level when minimizing side-effects has less importance. You will also notice the strength of the drug is much higher, particularly at the beginning of the treatment period. However, what is perhaps most interesting is that with less significance placed on side-effects, the same general strategy of chemotherapy should be employed, namely, a very high initial strength followed by a gradual reduction to no drug treatment.

Run the program with the (10.1) parameters, this time varying with $a = 1$. We have two systems, one where minimizing tumor density is three times as important as minimizing drug side-effects, and the second where they are of equal importance. The results are as we would expect. In the first system, a stronger drug regimen is used, reducing the tumor density to lower levels.

The previous simulations were run with initial tumor density near carrying capacity. Try (10.1) varying N_0 now to something smaller, say $N_0 = 0.5$. Notice how the two tumor densities and drug strengths converge. By day 8, the two systems are nearly identical. It seems only the early stages of optimal treatment are affected by initial tumor density. Afterwards, treatment and results become uniform. Now try the two initial densities of $N_0 = 0.975$ and $N_0 = 0.3$. Even in this more extreme case, virtually the same thing happens, at almost the same rate. Next, use the same two N_0 values with a longer time interval. Namely, try

$$\boxed{r = 0.3 \quad a = 3 \quad \delta = 0.45 \quad N_0 = 0.975 \quad T = 40} \tag{10.2}$$

varying with $N_0 = 0.3$. Again, the same convergence occurs. However, instead of being scaled to the new interval, uniformity still occurs in approximately 8 days.

Moving to the growth rate of the tumor, enter

$$\boxed{r = 0.3 \quad a = 3 \quad \delta = 0.45 \quad N_0 = 0.6 \quad T = 20}. \tag{10.3}$$

Vary the growth rate using the second value $r = 0.5$. As expected, the higher growth rate in the second system causes the tumor density to decrease more slowly. Also, the overall pharmokinetics in the second system must be greater to compensate. However, note that the drug strength in the first system begins at a higher level, before falling below the second control. With a slower growth rate, the initial blitz of drug is even more effective, so more is used.

Let us now consider the magnitude of the dosage δ. Enter the values

$$\boxed{r = 0.2 \quad a = 3 \quad \delta = 0.25 \quad N_0 = 0.8 \quad T = 20}. \tag{10.4}$$

Vary the magnitude using the second value $\delta = 0.5$. Even with a higher dose magnitude, the second system has an optimal drug strength which begins higher than the first. It then experiences a much faster reduction. The difference in tumor densities is fairly dramatic compared to our earlier simulations. This is the strongest evidence we have seen of the disproportional importance of drug effect in the first few days. In this example, the drug strength in the second system is slightly higher early, which, along with a higher dosage magnitude, drives the tumor density down. By day 6, both drug regimens have been lowered enough so that tumor density is being held approximately constant. However, over the 20 day period, the tumor density in the second system is much lower, almost half during much of the time. This difference is created almost entirely in the first 4 days.

Finally, examine the effect of the number of days on the optimal treatment. Enter

$$\boxed{r = 0.3 \quad a = 1.5 \quad \delta = 0.5 \quad N_0 = 0.7 \quad T = 20}. \tag{10.5}$$

Vary the number of days using the second value $T = 40$. We see the second system uses the same basic strategy as the first system, stretched over the longer interval. In fact, the strategies are almost identical for the first 10 days. The second system holds the mid-level drug strength for the next 20 days before dropping down, just as the first system did in days 10 through 20.

You may have noticed that in each simulation, the bound on u seemed unnecessary. The optimal control in each case was smooth and everywhere non-negative. It never appeared to be "cut-off" at zero. In fact, one can prove that the bound is superfluous; the optimal control will be non-negative without the bound, for accepted parameter values. However, it is standard practice in most research to include any relevant bounds, whether they are actually required or not. As you will see, the next lab contains the same bound, which is necessary there.

Exercise 10.1 Consider the ODE without control

$$N'(t) = rN(t) \ln\left(\frac{1}{N(t)}\right), \ N(0) = N_0.$$

Show that for $0 < N_0 < 1$, we have $N' > 0$ so that N is increasing. What happens when $N_0 = 1$?

Exercise 10.2 Prove analytically that the lower bound $u \geq 0$ in the optimal control is unnecessary. Namely, show that an optimal control for the problem

$$\min_u \int_0^T aN(t)^2 + u(t)^2 \, dt$$

subject to $\quad N'(t) = rN(t) \ln\left(\frac{1}{N(t)}\right) - u(t)\delta N(t), \ N(0) = N_0$

satisfies $u^* \geq 0$.

Chapter 11

Lab 6: Fish Harvesting

In this lab, we examine a simple fish harvesting problem adapted from [68]. Suppose at some point, designated as $t = 0$, a fish population is introduced into a fishery of some kind (for example, an artificial tank or a netted area in a body of water). Let $x(t)$ be the population level (scaled) at time t, where $x(0) = x_0 > 0$ is the initial concentration, as determined by the introduction. Suppose that, when introduced, the fish are very small and that the average mass of the fish at time $t = 0$ is essentially 0. Further, the average mass of the fish as a function of time is given by

$$f_{mass}(t) = \frac{kt}{t+1},$$

where k is the maximum mass of this species. We will assume the time interval $[0, T]$, over which we are to consider harvesting, is small enough that no reproduction will occur. Specifically, the population will have no natural growth. Let $u(t)$ be the harvest rate at time t and m be the natural death rate of the fish. We wish to maximize the total mass harvested over the interval taking into account the cost of harvesting. So, the problem can be stated

$$\max_u \int_0^T A \frac{kt}{t+1} x(t) u(t) - u(t)^2 \, dt$$

$$\text{subject to} \quad x'(t) = -(m + u(t))x(t), \quad x(0) = x_0,$$
$$0 \le u(t) \le M.$$

The upper bound M is added to take physical limitations of harvesting into account, and A is a nonnegative weight parameter. Note, if u is set to 0, then $x(t) = x_0 e^{-mt}$ which naturally decreases. Any positive control will cause the state to decrease even more.

Type *lab6* and press enter. Enter the values

$$\boxed{A = 5 \quad k = 10 \quad m = 0.2 \quad x_0 = 0.4 \quad M = 1 \quad T = 10}. \tag{11.1}$$

Do not vary any parameters. For reference, a plot of the average mass is also displayed. The optimal harvesting strategy here is no harvesting early, followed by a sharp increase, then a slow reduction. You can see the fish population begins a more rapid decrease instantly when harvesting begins.

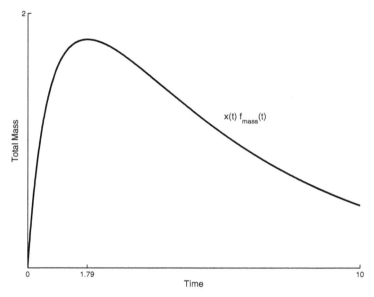

FIGURE 11.1: Graph of total mass, population x times average mass f_{mass}, with no control, for the values (11.1). The graph reaches its maximum at $t = 1.79$.

Also, the fish population is almost totally wiped out during our period. Notice, the average mass reaches 9 units in our 10 day simulation. Harvesting begins when the average mass is approximately 4.5. Figure 11.1 displays the state x, where no control is used, times the average mass f_{mass}.

The upper bound used in this simulation had no effect, as it was set relatively high. Enter the (11.1) values, varying with $M = 0.6$. The second control is approximately a truncation of the first, but not exactly. There is little change in the states. Enter (11.1) and vary with $M = 0.4$. The fact that the second control is not a truncation of the first is more apparent here. Also, the states are more distinct. We will refrain from exploring the effect of the upper bound any further here, as this was done in Lab 4.

Alter the weight parameter A. Enter (11.1), varying with $A = 10$. With a higher importance placed on the harvested weight, we expect a stronger control. In this simulation, the second control, while stronger overall, is not everywhere greater than the first control. The second control remains at no harvest for a longer period at the beginning, but eventually reaches a higher level. When total harvesting is more important, we should allow the fish to become larger, then harvest a larger percentage.

Experimenting with the maximum mass k, you would find its effect on the control is precisely the same as that of A. Even though they have very different meanings physically, these two constants affect the control the same

Lab 6: Fish Harvesting

mathematically, because they appear in identical places in the problem. In fact, if you ran two simulations, one with $A = 10$ and $k = 5$, and the other with $A = 5$ and $k = 10$, you would find the controls are the same. It is the product Ak which affects the control.

Now vary the death rate m. Enter the (11.1) values, varying with $m = 0.4$. With a higher natural death rate, the resulting optimal control begins harvesting sooner. It reaches a higher maximum harvesting level, but decreases faster as well. As more fish are going to die naturally, we should harvest more upfront. The state in the second system is everywhere less than the first, due to the change in death and harvesting rates.

Enter (11.1) and vary with $x_0 = 0.8$. Here, the new control begins harvesting later, but reaches a greater level. In fact, after passing the first control around $t = 1$, the second harvesting rate is larger for the remainder of the time period. Because the second fish population has a larger initial size, we are able to neglect the death rate a little longer before beginning harvesting. Consequently, the fish have a higher average mass, and the control can be increased, as the larger yield will offset the boost in harvesting rate.

Finally, vary the length of the time period. Enter (11.1), comparing $T = 10$ to $T = 5$. The optimal control and corresponding states are relatively similar over their common time. Now try $T = 10$ vs. $T = 15$ and $T = 10$ vs. $T = 20$. The two systems in these simulations have even greater similarity. This is fairly unusual. As we have seen in previous labs, and will continue to see in later ones, altering the time period usually affects the dynamics of the optimal control and state. Explain why this might occur here.

Chapter 12

Optimal Control of Several Variables

So far, we have only examined problems with one control and one dependent state variable. Often, though, we will wish to consider more variables. For example, consider a system modeling antibiotics used to fight a viral infection. In addition to the number of viral particles in the blood, we might also want to follow the number of antibodies or white blood cells. These quantities would be represented as additional state variables. Further, suppose the patient was taking two different antibiotics that caused the body to generate antibodies at different rates or times. These would need to be separate control variables; see [90]. Further, we could examine an SIR epidemic model with vaccination levels as a control [13, 93, 137, 149, 157, 180], or a tuberculosis epidemic model involving decisions in allocating efforts [95]. In this chapter, we will discuss how to handle such problems.

12.1 Necessary Conditions

The methods developed for one control and state are easily extended to optimal control of multiple state and control variables. Consider a problem with n state variables, m control variables, and a payoff function ϕ,

$$\max_{u_1,\ldots,u_m} \int_{t_0}^{t_1} f(t, x_1(t), \ldots, x_n(t), u_1(t), \ldots, u_m(t))\, dt + \phi(x_1(t_1), \ldots, x_n(t_1))$$

$$\text{subject to}\quad x_i'(t) = g_i(t, x_1(t), \ldots, x_n(t), u_1(t), \ldots, u_m(t)),$$
$$x_i(t_0) = x_{i0} \text{ for } i = 1, 2, \ldots, n,$$

where the functions f, g_i are continuously differentiable in all variables. We make no requirements on m, n. In fact, $m < n$, $m = n$, or $m > n$ are all acceptable. Use vector notation to change the problem to a more familiar form. Let $\vec{x}(t) = [x_1(t), \ldots, x_n(t)]$, $\vec{u}(t) = [u_1(t), \ldots, u_m(t)]$, $\vec{x}_0 = [x_{10}, \ldots, x_{n0}]$, and $\vec{g}(t, \vec{x}, \vec{u}) = [g_1(t, \vec{x}, \vec{u}), \ldots, g_n(t, \vec{x}, \vec{u})]$. Then, we can write the problem as

$$\max_{\vec{u}} \int_{t_0}^{t_1} f(t, \vec{x}(t), \vec{u}(t))\, dt + \phi(\vec{x}(t_1))$$

subject to $\vec{x}'(t) = \vec{g}(t, \vec{x}(t), \vec{u}(t))$, $\vec{x}(t_0) = \vec{x}_0$.

Let $\vec{u}^*$ be a vector of optimal control functions and $\vec{x}^*$ be the vector of corresponding optimal state variables. With n states, we will need n adjoints, one for each state. Introduce a piecewise differentiable vector-valued function $\vec{\lambda}(t) = [\lambda_1(t), ..., \lambda_n(t)]$, where each λ_i is the adjoint variable corresponding to x_i. Define the Hamiltonian

$$H(t, \vec{x}, \vec{u}, \vec{\lambda}) = f(t, \vec{x}, \vec{u}) + \vec{\lambda}(t) \cdot \vec{g}(t, \vec{x}, \vec{u}),$$

where $\cdot$ is the dot product of vectors. By essentially the same argument presented in Chapter 1, we find the variables satisfy identical optimality, adjoint, and transversality conditions in each vector component. Namely, $\vec{u}^*$ maximizes $H(t, \vec{x}^*, \vec{u}, \vec{\lambda})$ with respect to $\vec{u}$ at each t, and $\vec{u}^*$, $\vec{x}^*$, and $\vec{\lambda}$ satisfy

$$x'_i(t) = \frac{\partial H}{\partial \lambda_i} = g_i(t, \vec{x}, \vec{u}), \ x_i(t_0) = x_{i0} \text{ for } i = 1, ..., n,$$

$$\lambda'_j(t) = -\frac{\partial H}{\partial x_j}, \ \lambda_j(t_1) = \phi_{x_j}(\vec{x}(t_1)) \text{ for } j = 1, ..., n,$$

$$0 = \frac{\partial H}{\partial u_k} \text{ at } u_k^* \text{ for } k = 1, ..., m,$$

where

$$H(t, \vec{x}, \vec{u}, \vec{\lambda}) = f(t, \vec{x}, \vec{u}) + \sum_{i=1}^{n} \lambda_i(t) g_i(t, \vec{x}, \vec{u}).$$

By ϕ_{x_j}, it is meant the partial derivative in the x_j component. Note, if $\phi \equiv 0$, then $\lambda_j(t_1) = 0$ for all j, as usual.

Modifications of the problems yield adjustments on the conditions similar to those in previous chapters. For example, if a particular state variable x_i satisfies $x_i(t_0) = x_{i0}$, $x_i(t_1) = x_{i1}$ both fixed, then the corresponding adjoint λ_i has no boundary conditions. Similarly, if bounds are placed on a control variable, $a_k \leq u_k \leq b_k$, then the optimality condition is changed from

$$\frac{\partial H}{\partial u_k} = 0 \quad \text{to} \quad \begin{cases} u_k = a_k & \text{if } \frac{\partial H}{\partial u_k} < 0, \\ a_k \leq u_k \leq b_k & \text{if } \frac{\partial H}{\partial u_k} = 0, \\ u_k = b_k & \text{if } \frac{\partial H}{\partial u_k} > 0. \end{cases}$$

We illustrate these ideas with a few examples.

Example 12.1

$$\min_u \int_0^1 x_2(t) + u(t)^2 \, dt$$

subject to $x_1'(t) = x_2(t), \ x_1(0) = 0, \ x_1(1) = 1,$
$x_2'(t) = u(t), \ x_2(0) = 0.$

Introduce two adjoint variables, one for each state variable, and form the Hamiltonian,

$$H = x_2 + u^2 + \lambda_1 x_2 + \lambda_2 u.$$

Form the adjoint and transversality conditions

$$\lambda_1'(t) = -\frac{\partial H}{\partial x_1} = 0,$$
$$\lambda_2'(t) = -\frac{\partial H}{\partial x_2} = -\lambda_1 - 1, \ \lambda_2(1) = 0.$$

The first adjoint λ_1 is simply a constant, say C. Then, λ_2 can be solved as follows,

$$\lambda_1(t) \equiv C,$$
$$\lambda_2(t) = -(C+1)(t-1).$$

Using the optimality condition,

$$0 = \frac{\partial H}{\partial u} = 2u + \lambda_2 \ \Rightarrow \ u^* = -\frac{\lambda_2}{2} = \frac{C+1}{2}(t-1).$$

Finally, we make use of the state equations and boundary conditions to find

$$x_2' = u \Rightarrow x_2(t) = \frac{C+1}{2}\left(\frac{t^2}{2} - t\right), \text{ as } x_2(0) = 0,$$
$$x_1' = x_2 \Rightarrow x_1(t) = \frac{C+1}{2}\left(\frac{t^3}{6} - \frac{t^2}{2}\right), \text{ as } x_1(0) = 0.$$

Noting that $x_1(1) = 1$, it follows $C = -7$. Thus, the optimal solution set is

$$u^*(t) = 3 - 3t, \quad x_1^*(t) = \frac{3}{2}t^2 - \frac{1}{2}t^3, \quad x_2^*(t) = 3t - \frac{3}{2}t^2.$$

The optimal states x_1^* and x_2^* are shown in Figure 12.1.

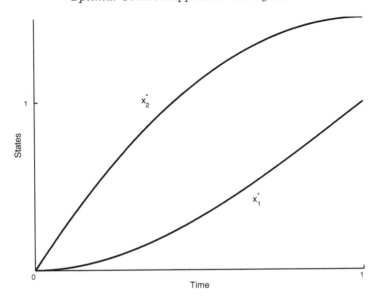

FIGURE 12.1: Optimal states for Example 12.1.

Example 12.2

$$\max_{u_1,u_2} \int_0^1 x(t) - \frac{1}{8}u_1(t)^2 - \frac{1}{2}u_2(t)^2 \, dt$$

$$\text{subject to} \quad x'(t) = u_1(t) + u_2(t), \ x(0) = 0,$$
$$1 \leq u_1(t) \leq 2.$$

The Hamiltonian is

$$H = x - \frac{1}{8}u_1^2 - \frac{1}{2}u_2^2 + \lambda u_1 + \lambda u_2.$$

The adjoint and transversality conditions yield

$$\lambda'(t) = -\frac{\partial H}{\partial x} = -1, \ \lambda(1) = 0 \Rightarrow \lambda(t) = 1 - t.$$

The second control has no bounds, so we can easily solve for it,

$$0 = \frac{\partial H}{\partial u_2} = -u_2 + \lambda \Rightarrow u_2^* = \lambda = 1 - t.$$

To find u_1^*, we note

$$\frac{\partial H}{\partial u_1} = \lambda - \frac{u_1}{4},$$

$$\frac{\partial H}{\partial u_1} < 0 \Rightarrow u_1(t) = 1 \Rightarrow 1 - t < \frac{1}{4} \Rightarrow t > \frac{3}{4},$$

$$\frac{\partial H}{\partial u_1} = 0 \Rightarrow u_1^* = 4\lambda = 4 - 4t \Rightarrow 1 \leq 4 - 4t \leq 2 \Rightarrow \frac{1}{2} \leq t \leq \frac{3}{4},$$

$$\frac{\partial H}{\partial u_1} > 0 \Rightarrow u_1(t) = 2 \Rightarrow 1 - t > \frac{1}{2} \Rightarrow t < \frac{1}{2}.$$

By plugging the three cases back into the state equation, and requiring continuity, we can find x^*. Then, the optimal solution set (Figure 12.2) is

$$u_1^*(t) = \begin{cases} 2 & 0 \leq t < \frac{1}{2}, \\ 4 - 4t & \frac{1}{2} \leq t \leq \frac{3}{4}, \\ 1 & \frac{3}{4} < t \leq 1, \end{cases} \qquad u_2^*(t) = 1 - t,$$

$$x^*(t) = \begin{cases} 3t - \frac{1}{2}t^2 & 0 \leq t \leq \frac{1}{2}, \\ 5t - \frac{5}{2}t^2 - \frac{1}{2} & \frac{1}{2} \leq t \leq \frac{3}{4}, \\ 2t - \frac{1}{2}t^2 + \frac{5}{8} & \frac{3}{4} \leq t \leq 1. \end{cases}$$

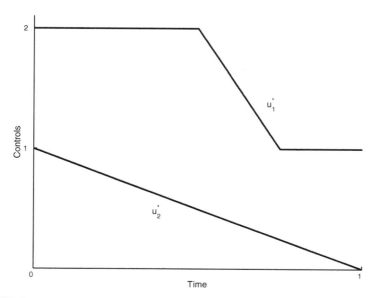

FIGURE 12.2: Optimal controls for Example 12.2.

With certain optimal control problems, it is sometimes possible to ignore some of the states when forming the necessary conditions. Reducing the number of states in the optimality system can simplify matters, particularly by also eliminating the corresponding adjoints. This technique is used in Labs 7 and 12, and we present two examples here.

Example 12.3 Consider the following general optimal control problem

$$\max_u \int_{t_0}^{t_1} f(t, x_1(t), x_2(t), u(t))\, dt$$

subject to $\quad x_1'(t) = g_1(t, x_1(t), x_2(t), u(t)), \quad x_1(t_0) = x_{10},$
$\qquad\qquad\;\; x_2'(t) = g_2(t, x_1(t), x_2(t), u(t)), \quad x_2(t_0) = x_{20},$
$\qquad\qquad\;\; x_3'(t) = g_3(t, x_1(t), x_2(t), x_3(t), u(t)), \quad x_3(t_0) = x_{30}.$

Note, f, g_1, and g_2 have no dependence on the state variable x_3. Therefore,

$$\max_u \int_{t_0}^{t_1} f(t, x_1(t), x_2(t), u(t))\, dt$$

subject to $\quad x_1'(t) = g_1(t, x_1(t), x_2(t), u(t)), \quad x_1(t_0) = x_{10},$
$\qquad\qquad\;\; x_2'(t) = g_2(t, x_1(t), x_2(t), u(t)), \quad x_2(t_0) = x_{20},$

is a valid, well-posed optimal control problem, which can be solved as normal. Once $x_1^*(t)$, $x_2^*(t)$, and $u^*(t)$ are found, then $x_3^*(t)$ can be solved using its state equation and initial condition.

In the next example, we are able to eliminate a state variable by manipulating the objective functional.

Example 12.4 (from [82]) Consider a well-stirred bioreactor in which contaminant and bacteria are present in spatially uniform, time varying concentrations:

$$z(t) = \text{concentration of contaminant,}$$
$$x(t) = \text{concentration of bacteria.}$$

The bioreactor is rich in all nutrients except one to be controlled,

$$u(t) = \text{concentration of input nutrient.}$$

The bacteria degrades the contaminant via co-metabolism, meaning degradation of the contaminant is a byproduct of the bacteria metabolism; the bacteria does not consume the contaminant directly. The growth of the bacteria as result of the nutrient u is a limited growth function, called Monod or Michealis-Menton kinetics:

$$x'(t) = \frac{Gu(t)}{H+u(t)} x(t) - Dx(t)^2, \ x(0) = x_0,$$
$$z'(t) = -Kx(t)z(t), \ z(0) = z_0.$$

Here, G is maximum growth rate, D is the death rate, and K is the degradation rate of the bacteria. The constant H is the nutrient concentration at which the bacterial growth rate is one half its maximum value. We wish to minimize the final contaminant concentration and total injection of the nutrient. As the natural logarithm is a strictly increasing function, we can minimize the objective functional

$$\ln(z(T)) + \int_0^T Au(t)\,dt.$$

The introduction of the natural logarithm may seem artificial, but it allows a great simplification. Note, also, the first state variable x has no dependence on z in its state equation. So, if we can find an expression of $z(T)$ in terms of t, x, and/or u, we could eliminate the variable z from the problem. Solving the equation

$$z'(t) = -Kx(t)z(t),$$

it follows that

$$z(t) = z_0 \exp\left(-\int_0^t Kx(s)\,ds\right).$$

Using this expression with $t = T$,

$$\int_0^T Kx(s)\,ds = -\ln\left(\frac{z(T)}{z_0}\right) = \ln(z_0) - \ln(z(T)).$$

The constant $\ln(z_0)$ is irrelevant to the minimization, so we ignore it. Replacing $\ln(z(T))$ with the integral expression above, and negating, our problem can be cast as

$$\max_u \int_0^T Kx(t) - Au(t)\,dt$$

$$\text{subject to} \quad x'(t) = \frac{Gu(t)}{H+u(t)} x(t) - Dx(t)^2, \ x(0) = x_0,$$
$$u(t) \geq 0.$$

We have eliminated the z variable, so there is no need to include its state equation in the problem. Once we have found the optimal control and x^*, we can then find z^*. For now, we work only with x, and thus, only one adjoint. Using the Hamiltonian, we calculate the necessary conditions.

$$H = Kx - Au + \lambda\left(\frac{Gux}{H+u} - Dx^2\right),$$

$$\frac{\partial H}{\partial u} = -A + \lambda x \frac{GH}{(H+u)^2} \Rightarrow u^* = \max\left(0, A^{-1/2}(\lambda x GH)^{1/2} - H\right),$$

$$\lambda' = -\frac{\partial H}{\partial x} = -\left[\lambda\left(\frac{Gu}{H+u} - 2Dx\right) + K\right], \quad \lambda(T) = 0.$$

We could now solve for u^*, x^*, and λ numerically. A bioreactor problem with a simpler growth term is the subject of Lab 12.

12.2 Linear Quadratic Regulator Problems

In this section, we treat a special case in the optimal control of systems, in which the state differential equations are linear in x and u and the objective functional is quadratic. A solution can be found in a slightly different way in this case and has a very nice format. In particular, we are able to eliminate the adjoint variable in the necessary conditions. For example, one might use such systems to model chemostats [170].

Our state system is given by

$$x'(t) = A(t)x(t) + B(t)u(t), \quad (12.1)$$

where x is an n-dimensional column vector, and u is a m-dimensional column vector. The matrices $A(t), B(t)$ have sizes $n \times n$ and $n \times m$ respectively. Note that entries of matrices of A, B can be functions of time. The objective functional is

$$J(u) = \frac{1}{2}\left[x^\mathsf{T}(T)Mx(T) + \int_0^T x^\mathsf{T}(t)Q(t)x(t) + u^\mathsf{T}(t)R(t)u(t)\,dt\right], \quad (12.2)$$

where the symmetric matrices M, $Q(t)$, and $R(t)$ are sizes $n \times n$, $n \times n$, and $m \times m$ respectively, with M, $Q(t)$ being positive semidefinite and $R(t)$ being positive definite for all $0 \le t \le T$. The positive definite property guarantees $R(t)$ is invertible. The superscript T refers to transpose of the matrix. We can interpret the objective functional as minimizing a weighted sum of the components of the state and the control. The matrices would be chosen to decide which components to emphasize. In practice, the state might be the difference between a quantity (like the levels of microorganisms in a chemostat) and its desired profile, and the objective functional can drive certain components of the quantity close to the profile.

Like the control and state, we write λ to mean an n-dimensional column vector of adjoints. The Hamiltonian becomes

$$H = \frac{1}{2}x^T Q x + \frac{1}{2}u^T R u + \lambda^T(Ax + Bu).$$

Some care must be taken in differentiating matrix expressions, particularly if not familiar with the process. We suppress the details here, but encourage the reader to check the calculations term-by-term. The optimality equation is

$$Ru + B^T\lambda = 0 \Rightarrow u^* = -R^{-1}B^T\lambda,$$

and the adjoint equation is

$$\lambda' = -Qx - A^T\lambda, \ \lambda(T) = Mx(T).$$

The assumptions of symmetry for M, Q, and R are buried in the above calculations. We choose to solve this problem in a different way due to the structure of the transversality condition and the adjoint differential equation; this method is called the *sweep method* [20, 51, 126]. Instead of using λ, we find a matrix function $S(t)$ such that $\lambda(t) = S(t)x(t)$. By the product rule for matrices,

$$\lambda'(t) = S'(t)x(t) + S(t)x'(t).$$

Using the expressions for λ' and x' given by the state and adjoint equations, we have

$$-Qx - A^T\lambda = S'x + SAx + SBu.$$

Making use of the characterization of the control and the identity $\lambda = Sx$,

$$-S'x = Qx + A^T\lambda + SAx - SBR^{-1}B^T\lambda$$
$$= Qx + A^T Sx + SAx - SBR^{-1}B^T Sx$$
$$= [Q + A^T S + SA - SBR^{-1}B^T S]x.$$

From the transversality condition, we obtain the matrix *Riccati* equation that $S(t)$ must satisfy. Namely,

$$-S' = A^T S + SA - SBR^{-1}B^T S + Q, \ S(T) = M.$$

Reconsidering the characterization, we see the control is a linear function of the state only, a type of feedback control

$$u = -R^{-1}B^\mathsf{T} Sx.$$

The matrix $R^{-1}B^\mathsf{T} S$ is called the *gain*. After solving the Riccati matrix equation for S, the control is given by an equation in x, and x is given by an ODE in u, so that the problem can be solved using simple ODE methods. Therefore, we have totally eliminated the adjoint λ from the problem. See the book by Morris about feedback control [148] and a recent application of the Riccati approach [9].

Example 12.5 (from [51]) We consider a simple one dimensional example.

$$\frac{1}{2} \min_u \int_0^T x(t)^2 + u(t)^2 \, dt$$

subject to $x'(t) = u(t),\ x(0) = x_0.$

In this case, all the matrices are scalars (size 1×1) and $S(T) = M = 0$, $A = 0,\ B = Q = R = 1$. The Riccati equation is

$$-S' = 1 - S^2,\ S(T) = 0.$$

Solving as a separable equation, and using partial fractions,

$$\frac{1}{2} \ln \left| \frac{S-1}{S+1} \right| = \int \frac{S'}{S^2 - 1} \, dt = \int 1 \, dt = t + C,$$

which along with $S(T) = 0$ gives

$$S(t) = \frac{1 - e^{2(t-T)}}{1 + e^{2(t-T)}}.$$

The optimal control satisfies $u = -Sx$, so that the optimal state satisfies $x' = -Sx$. Using partial fractions (or an integral table) we can find an antiderivative of S, and solve the separable equation to see

$$x(t) = C(e^{t-T} + e^{T-t}).$$

Taking into account $x(0) = x_0$,

$$x^*(t) = x_0 \frac{e^t + e^{2T-t}}{1 + e^{2T}}, \quad \text{and} \quad u^*(t) = x_0 \frac{e^t - e^{2T-t}}{1 + e^{2T}}.$$

12.3 Higher Order Differential Equations

Optimal control of systems can be employed to solve maximization (or minimization) problems involving higher order differential equations. Consider the following problem,

$$\max_{u_1,\ldots,u_m} \int_{t_0}^{t_1} f(t, x(t), x'(t), \ldots, x^{(n)}(t), u_1(t), \ldots, u_m(t))\, dt$$

subject to $x^{(n+1)}(t) = g(t, x(t), x'(t), \ldots, x^{(n)}(t), u_1(t), \ldots, u_m(t))$,

$$x(t_0) = \alpha_1,\ x'(t_0) = \alpha_2,\ \ldots,\ x^{(n)}(t_0) = \alpha_{n+1}$$

for $n > 1$. Pontryagin's Maximum Principle, as we have developed it, does not directly deal with this type of problem. However, it is easily converted to a systems problem by introducing $n+1$ state variables defined by $x_1(t) = x(t)$, $x_2(t) = x'(t)$, $\ldots$, $x_{n+1}(t) = x^{(n)}(t)$. Then, the above problem becomes

$$\max_{u_1,\ldots,u_m} \int_{t_0}^{t_1} f(t, x_1(t), x_2(t), \ldots, x_{n+1}(t), u_1(t), \ldots, u_m(t))\, dt$$

subject to $x_1'(t) = x_2(t),\ x_1(t_0) = \alpha_1,$
$\qquad\qquad x_2'(t) = x_3(t),\ x_2(t_0) = \alpha_2,$

$$\vdots$$

$\qquad\qquad x_n'(t) = x_{n+1}(t),\ x_n(t_0) = \alpha_n,$
$\qquad\qquad x_{n+1}'(t) = g(t, x_1(t), x_2(t), \ldots, x_{n+1}(t), u_1(t), \ldots, u_m(t)),$
$\qquad\qquad x_{n+1}(t_0) = \alpha_{n+1}$

which can be solved by the methods developed in this chapter. Consider this second-order example.

Example 12.6

$$\min_u \frac{1}{2} \int_0^\pi u(t)^2 - x(t)^2\, dt$$

subject to $x''(t) = u(t),\ x(0) = 1,\ x'(0) = 1.$

Let $x_1 = x$ and $x_2 = x'$, to convert the problem to

$$\min_u \frac{1}{2} \int_0^\pi u(t)^2 - x_1(t)^2\, dt$$

subject to $x_1'(t) = x_2(t),\ x_1(0) = 1,$
$\qquad\qquad x_2'(t) = u(t),\ x_2(0) = 1.$

Introduce two adjoint variables λ_1 and λ_2 and set up the Hamiltonian:

$$H = \frac{1}{2}u^2 - \frac{1}{2}x_1^2 + \lambda_1 x_2 + \lambda_2 u,$$

$$0 = \frac{\partial H}{\partial u} = u + \lambda_2 \Rightarrow u^* = -\lambda_2,$$

$$\lambda_1'(t) = -\frac{\partial H}{\partial x_1} = x_1, \; \lambda_1(\pi) = 0,$$

$$\lambda_2'(t) = -\frac{\partial H}{\partial x_2} = -\lambda_1, \; \lambda_2(\pi) = 0.$$

Note, $\lambda_2^{(4)} = -\lambda_1''' = -x_1'' = -x_2' = -u = \lambda_2$. Thus, $\lambda_2(t) = Ae^t + Be^{-t} + C\cos t + D\sin t$ for some constants A, B, C, D. Making use of the adjoint and state equations, we see

$$x_1(t) = -Ae^t - Be^{-t} + C\cos t + D\sin t,$$
$$x_2(t) = -Ae^t + Be^{-t} - C\sin t + D\cos t,$$
$$\lambda_1(t) = -Ae^t + Be^{-t} + C\sin t - D\cos t,$$
$$\lambda_2(t) = Ae^t + Be^{-t} + C\cos t + D\sin t.$$

Using the conditions $x_1(0) = x_2(0) = 1$ and $\lambda_1(\pi) = \lambda_2(\pi) = 0$, we find

$$\begin{pmatrix} -1 & -1 & 1 & 0 \\ -1 & 1 & 0 & 1 \\ -e^\pi & e^{-\pi} & 0 & 1 \\ e^\pi & e^{-\pi} & -1 & 0 \end{pmatrix} \begin{pmatrix} A \\ B \\ C \\ D \end{pmatrix} = \begin{pmatrix} 1 \\ 1 \\ 0 \\ 0 \end{pmatrix}.$$

Approximate values are $A \approx 0.0452$, $B = 0$, $C = D \approx 1.0452$, so that the optimal solutions are

$$x^*(t) = x_1(t) = -0.0452e^t + 1.0452(\cos(t) + \sin(t)),$$
$$u^*(t) = -\lambda_2(t) = -0.0452e^t - 1.0452(\cos(t) + \sin(t)).$$

12.4 Isoperimetric Constraints

Let us return our attention to Example 3.3, the simple problem involving cancer treatment:

$$\min_u \; x(T) + \int_0^T u(t)^2 \, dt$$

subject to $\quad x'(t) = \alpha x(t) - u(t), \ x(0) = x_0.$

As originally stated, we wished to minimize the final concentration of tumor cells and the total harmful effects of the drugs. Now, suppose we wanted to further restrict the amount of drug administered to the patient. One method would be to introduce a bound on the control, say $0 \leq u(t) \leq M$, where M is an appropriately chosen constant. This method still allows some leniency in the total amount of drug. Suppose we know precisely the amount of treatment which can be given to this patient over the given time interval and still be within safety limits. Further, suppose we wish to administer precisely this amount over the time period, or stated mathematically

$$\int_0^T u(t)\,dt = B,$$

where B is the known amount. Then, the problem we are now faced with is to minimize the final concentration of cancerous cells using a total drug amount of B over the time interval. This can be stated

$$\min_u \ x(T) + \int_0^T u(t)^2\,dt$$

subject to $\quad x'(t) = \alpha x(t) - u(t), \ x(0) = x_0,$

$$\int_0^T u(t)\,dt = B.$$

This type of constraint is known as an *isoperimetric constraint*. We now proceed to establish solution methods for this type of problem in more generality.

Let $f(t,x,u)$, $g(t,x,u)$, and $h(t,x,u)$ be continuously differentiable functions in all three variables. Consider the optimal control problem

$$\max_u \int_{t_0}^{t_1} f(t,x(t),u(t))\,dt + \phi(x(t_1))$$

subject to $\quad x'(t) = g(t,x(t),u(t)), \ x(t_0) = x_0,$

$$\int_{t_0}^{t_1} h(t,x(t),u(t))\,dt = B, \quad (12.3)$$

$$a \leq u(t) \leq b.$$

Pontryagin's Maximum Principle cannot be used to deal with this problem as stated. As in the last section, though, we can use a simple trick to convert this problem to a more familiar form. Introduce a second state variable $z(t)$ and set

$$z(t) = \int_{t_0}^t h(s,x(s),u(s))\,ds.$$

Then, it follows

$$z'(t) = h(t, x(t), u(t)),$$
$$z(t_0) = 0,$$
$$z(t_1) = B.$$

Thus, (12.3) is transformed into

$$\max_u \int_{t_0}^{t_1} f(t, x(t), u(t))\, dt + \phi(x(t_1))$$

subject to $\quad x'(t) = g(t, x(t), u(t)),\ x(t_0) = x_0,$
$\qquad\qquad\ \ z'(t) = h(t, x(t), u(t)),\ z(t_0) = 0,\ z(t_1) = B,$
$\qquad\qquad\ \ a \leq u(t) \leq b.$

This problem can now be solved using methods developed in this chapter. We present the following example. For an additional example, see Exercise 12.9.

Example 12.7

$$\min_u \frac{1}{2} \int_0^1 u(t)^2\, dt$$

subject to $\quad x'(t) = u(t),\ x(0) = 0,\ x(1) = 1,$

$$\int_0^1 x(t)\, dt = 2.$$

Introduce a second state variable $z(t)$ with $z'(t) = x(t)$, $z(0) = 0$, and $z(1) = 2$. Then, the above problem converts to

$$\min_u \frac{1}{2} \int_0^1 u(t)^2\, dt$$

subject to $\quad x'(t) = u(t),\ x(0) = 0,\ x(1) = 1,$
$\qquad\qquad\ \ z'(t) = x(t),\ z(0) = 0,\ z(1) = 2.$

The Hamiltonian will be

$$H = \frac{1}{2}u^2 + \lambda_1 u + \lambda_2 x.$$

The second adjoint equation is

$$\lambda_2'(t) = -\frac{\partial H}{\partial z} = 0,$$

so that $\lambda_2 \equiv C$ for some constant C. Also, the first adjoint equation yields

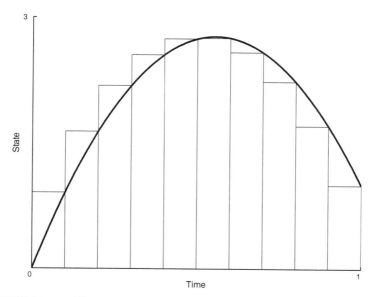

FIGURE 12.3: The optimal state from Example 12.7 is shown, along with rectangles for the right-hand sum approximation. Of course, here we have an expression we can integrate to check the isoperimetric constraint.

$$\lambda_1'(t) = -\frac{\partial H}{\partial x} = -\lambda_2 = -C \Rightarrow \lambda_1(t) = k - Ct$$

for some constant k. The optimality condition is

$$0 = \frac{\partial H}{\partial u} = u + \lambda_1 \Rightarrow u^*(t) = -\lambda_1(t) = Ct - k.$$

Using the state equation for x and $x(0) = 0$, we have

$$x'(t) = u(t) = Ct - k \Rightarrow x(t) = \frac{C}{2}t^2 - kt.$$

Similarly, using the state equation for z and $z(0) = 0$, we have

$$z'(t) = x(t) = \frac{C}{2}t^2 - kt \Rightarrow z(t) = \frac{C}{6}t^3 - \frac{k}{2}t^2.$$

Finally, using the terminal conditions,

$$1 = x(1) = \frac{C}{2} - k,$$
$$2 = z(1) = \frac{C}{6} - \frac{k}{2},$$

which can be solved to find $C = -18$ and $k = -10$. This gives the optimal solutions

$$u^*(t) = 10 - 18t,$$
$$x^*(t) = 10t - 9t^2.$$

The optimal state is shown in Figure 12.3.

12.5 Numerical Solutions

The forward-backward sweep method for several variables is essentially identical to the scheme used with one state and one control. First, an initial guess of each control variable is made. All states are solved simultaneously forward in time, then all adjoints are simultaneously solved backward in time. Each control is updated subject to its individual characterization, and the process is repeated until convergence occurs.

Any differential systems solver can be used to solve the states and adjoints. As Runge-Kutta 4 has been used to this point, we will now make use of Runge-Kutta 4 for systems, where if $\vec{x}'(t) = \vec{f}(t, \vec{x}(t))$, then

$$\vec{x}(t+h) = \vec{x}(t) + \frac{h}{6}(\vec{k}_1 + 2\vec{k}_2 + 2\vec{k}_2 + \vec{k}_4) \qquad (12.4)$$

where

$$\begin{aligned}
\vec{k}_1 &= \vec{f}(t, \vec{x}(t)) \\
\vec{k}_2 &= \vec{f}(t + \frac{1}{2}h, \vec{x}(t) + \frac{1}{2}h\vec{k}_1) \\
\vec{k}_3 &= \vec{f}(t + \frac{1}{2}h, \vec{x}(t) + \frac{1}{2}h\vec{k}_2) \\
\vec{k}_4 &= \vec{f}(t + h, \vec{x}(t) + h\vec{k}_3).
\end{aligned} \qquad (12.5)$$

Note that 12.4 and 12.5 are almost identical to the single equation Runge-Kutta 4 routine 4.1 and 4.2, except $\vec{x}$, $\vec{f}$, and $\vec{k}_i$, $i = 1, \ldots 4$, are now vectors, all of the same length.

To write a forward-backward sweep for an optimal control problem with multiple states and/or controls, make an initial guess for each control variable, declare all states and adjoints, and store all initial conditions. Also, when checking convergence before, we created a convergence measure for each of the control, state, and adjoint. Now, we must do the same for all controls, all states, and all adjoints, and then take the minimum. Each control will have its own characterization, which may or may not include bounds. In order to solve the states and adjoints, notice that the Runge-Kutta routine requires the vector $\vec{k}_i$ to be completely solved for before finding $\vec{k}_{i+1}$. This means the k_1 term for each state or adjoint must be found before moving onto the k_2 terms, and so on. Also, recall variables dependent on time which are not being solved for, such as the control, are shifted by taking an average.

For an example, suppose the state equations for an optimal control problem are as follows,

$$x_1'(t) = x_1 + x_2 u,$$
$$x_2'(t) = x_1 x_2 - u + t.$$

Then, the code would be written as below. Recall $h2 = h/2$.

```
for i = 1:N
    k11 = x1(i) + x2(i)*u(i);
    k12 = x1(i)*x2(i) - u(i) + t(i);

    k21 = (x1(i)+h2*k11) + (x2(i)+h2*k12)*0.5*(u(i)+u(i+1));
    k22 = (x1(i)+h2*k11)*(x2(i)+h2*k12) - ...
        0.5*(u(i)+u(i+1)) + (t(i) + h2);

    k31 = (x1(i)+h2*k21) + (x2(i)+h2*k22)*0.5*(u(i)+u(i+1));
    k32 = (x1(i)+h2*k21)*(x2(i)+h2*k22) - ...
        0.5*(u(i)+u(i+1)) + (t(i) + h2);

    k41 = (x1(i)+h*k31) + (x2(i)+h*k32)*u(i+1);
    k42 = (x1(i)+h*k31)*(x2(i)+h*k32) - u(i+1) + t(i+1);

    x1(i+1) = x1(i) + (h/6)*(k11 + 2*k21 + 2*k31 + k41);
    x2(i+1) = x2(i) + (h/6)*(k12 + 2*k22 + 2*k32 + k42);
end
```

The backward solver for the adjoints is similar.

12.6 Exercises

Exercise 12.1 (from [126]) Solve

$$\min_{u} 5x_1(1)^2 + \frac{1}{2}\int_0^1 x_1(t)^2 + x_2(t)^2 + u(t)^2 \, dt$$

subject to $x_1'(t) = x_2(t) + u(t),$
$x_2'(t) = x_1(t) - x_2(t), \ x_2(1) = 1.$

Exercise 12.2 Reconsider the previous exercise with the bounds $-1 \leq u(t) \leq 1$.

Exercise 12.3 Solve

$$\min_{u_1, u_2} \frac{1}{2}\int_0^1 u_1(t)^2 + u_2(t)^2 \, dt$$

subject to $x'(t) = u_1(t)u_2(t) + u_2(t), \ x(0) = 0, \ x(1) = 1.$

Exercise 12.4 Solve

$$\min_{u_1, u_2} \int_0^5 x_1(t) + \frac{1}{2}u_1(t)^2 + \frac{1}{2}u_2(t)^2 \, dt$$

subject to $x_1'(t) = u_1(t) + x_2(t), \ x_1(0) = x_0,$
$x_2'(t) = u_2(t), \ x_2(0) = 0,$
$-2 \leq u_1(t) \leq -1, \ -8 \leq u_2(t) \leq -2.$

Exercise 12.5 (from [51]) Consider the optimal control problem

$$\min_{u} \frac{1}{2}\int_0^1 x_1(t)^2 + u(t)^2 \, dt$$

subject to $x_1'(t) = x_2(t), \ x_1(0) = 1,$
$x_2'(t) = u(t), \ x_2(0) = 0.$

Cast this as an linear quadratic regulator problem. State the Ricatti equation.

Exercise 12.6 (from [148]) Cast as an LQR problem, and solve the Riccati equation.

$$\min_u \frac{1}{2} \int_0^T 3x(t)^2 + u(t)^2 \, dt$$

subject to $x'(t) = x(t) + u(t)$, $x(0) = x_0$.

Exercise 12.7 (from [148]) Cast as an LQR problem, and solve the Riccati equation.

$$\min_u \frac{1}{2} \int_0^T x_1(t)^2 + u(t)^2 \, dt$$

subject to $x_1'(t) = x_2(t)$, $x_1(0) = a$,

$x_2'(t) = -10x_2(t) + u(t)$, $x_2(0) = b$.

Exercise 12.8 Solve

$$\min_u \frac{1}{2} \int_0^1 u(t)^2 \, dt$$

subject to $x''(t) = u(t)$, $x(0) = 0$, $x(1) = A$, $x'(1) = 0$.

Exercise 12.9 Solve

$$\min_u \int_0^1 \frac{1}{2}x(t)^2 + x(t) + \frac{1}{2}u(t)^2 + u(t) \, dt$$

subject to $x'(t) = u(t)$, $x(0) = 0$,

$$\int_0^1 u(t) \, dt = 1.$$

Note, there is an alternate way of solving this problem. The isoperimetric constraint can be modified to a final time condition of the state using the Fundamental Theorem of Calculus.

Exercise 12.10 Formulate an optimal control problem for a system of three ordinary differential equations. This system represents three interacting populations. Population 1 and population 2 compete. Population 3 cooperates with the other two populations. Population 1 has logistic growth. The growth function of population 2 has an Allee effect. Population 3 has exponential growth. The control is to harvest a proportion of population 3. The objective functional should maximize the population harvested over time and minimize the cost of the harvest. Assume the cost of harvesting is a quadratic function of the control. Set-up the optimal control problem and the corresponding necessary conditions, but do not solve the conditions. (See Kot [107] for an explanation of the growth terms mentioned above.)

Chapter 13

Lab 7: Epidemic Model

In this lab, we use optimal control techniques to find a vaccination schedule for an epidemic disease. A micro-parasitic infectious disease is considered. Permanent immunity to the disease can be achieved through natural recovery or immunization. Immunity is not passed on during birth, so that everyone is born susceptible. Our goal is to minimize the number of infectious persons and the overall cost of the vaccine during a fixed time period.

To model the dynamics of the disease in a population, we use a standard SEIR (or SEIRN) model. Let $S(t)$, $I(t)$, and $R(t)$ represent number of susceptible, infectious, and recovered (immune) individuals at time t. The model allows for an incubation period for the disease inside its host, where an infected person remains latent for some time before becoming infectious, creating an exposed class. Let $E(t)$ be the number of exposed or latent individuals at time t. Let $N(t)$ be the total number of people in the population, so that $N(t) = S(t) + E(t) + I(t) + R(t)$.

Let $u(t)$, the control, be the percentage of susceptible individuals being vaccinated per unit of time. As vaccination of the entire susceptible population is impossible, we bound the control with $0 \leq u(t) \leq 0.9$. Let b be the natural exponential birth rate of the population and d the natural exponential death rate. The incidence of the disease is described by the term $cS(t)I(t)$. The parameter e is the rate at which the exposed individuals become infectious, and g is the rate at which infectious individuals recover. Therefore, $\frac{1}{e}$ is the mean latent period, and $\frac{1}{g}$ is the mean infectious period before recovery, if recovery occurs. The death rate due to the disease in infectious individuals is a. The optimal control problem is as follows,

$$\min_u \int_0^T AI(t) + u(t)^2 \, dt$$

subject to
$S'(t) = bN(t) - dS(t) - cS(t)I(t) - u(t)S(t), \ S(0) = S_0 \geq 0,$
$E'(t) = cS(t)I(t) - (e+d)E(t), \ E(0) = E_0 \geq 0,$
$I'(t) = eE(t) - (g+a+d)I(t), \ I(0) = I_0 \geq 0,$
$R'(t) = gI(t) - dR(i) + u(t)S(t), \ R(0) = R_0 \geq 0,$
$N'(t) = (b-d)N(t) - aI(t), \ N(0) = N_0,$
$0 \leq u(t) \leq 0.9.$

See [93] for results on a similar problem using incidence term $\frac{cSI}{N}$. The left-hand side of the differential equations gives us the name of the type of model (SEIR). A flow chart of the model is given in Figure 13.1. Also, observe the variable R appears only in the R' differential equation. So, the other variables do not depend on R, and we can ignore R when we solve the optimality system. Specifically, as you see in the code, only S, E, I, and N are solved forward in time, and the four associated adjoints are solved backward in time. Once convergence has been achieved, R^* is solved using its differential equation. Refer back to Example 12.3.

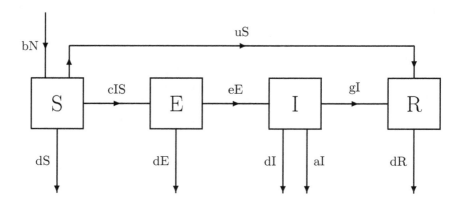

FIGURE 13.1: This is a flow chart for our model. The four boxes represent the four groups of individuals. The arrows show the movement between groups, and into and out of the population.

Type *lab7* at the prompt and press enter. Start with the values

$$\begin{array}{llll} b = 0.525 & d = 0.5 & c = 0.0001 & e = 0.5 \\ g = 0.1 & a = 0.2 & S_0 = 1000 & E_0 = 100 \\ I_0 = 50 & R_0 = 15 & A = 0.1 & T = 20 \end{array}. \quad (13.1)$$

This is a simulation of a disease with a low incidence measure. The optimal vaccination schedule is one of containment. An early round of vaccinations is used to shield the susceptible population from the initially significant exposed and infectious populations. This, combined with the low incidence level, results in the virtual end of disease spread. Exposed and infectious populations quickly disappear (through death and recovery). By year 5, the disease is essentially wiped out and vaccination ends. The small number of people who do carry the disease pose little threat of spreading it. The recovered group increases rapidly at first due to vaccinations, but slowly disappears when vaccination ends. By the end of the time period, susceptible people make up almost the entire population. Notice, the susceptible population decreases slightly at the beginning of the time interval. Here, the vaccination rate is greater than the overall growth. See Figure 13.2 to compare this simulation with that of no vaccination.

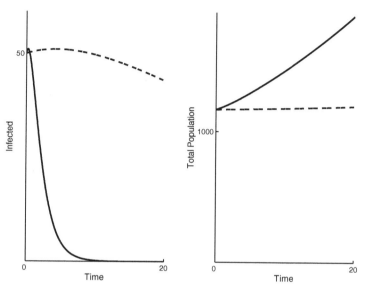

FIGURE 13.2: Results from optimal vaccination from (13.1), in solid, compared with that of no vaccination (dashed). Infectious I and total population N are pictured. With no vaccination, the number of infectious decreases little and total population is nearly constant.

Now, vary (13.1) with $c = 0.001$, a much higher, and more realistic, incidence level. Here, the threat of disease spread is much more serious, and a more aggressive plan is needed. Maximum vaccination is used initially, followed by a reduction, but a slower reduction than in the first system. With a low incidence, there was no need to vaccinate once the exposed and infectious populations were reduced, as almost no one would contract the disease. In the system with $c = 0.001$, though, we see it is advantageous to continue vaccinating almost 40% of the population, even after exposed and infectious populations are reduced. The susceptible population is reduced by half in the first two years, as the great majority are being vaccinated and many of the others are exposed. The infectious population even sees a slight initial increase. However, after the first several years, the same dynamic of the other simulation returns. Susceptible begins to steadily climb, almost reaching the levels of those in the first system. Exposed and infectious almost disappear, and vaccination levels do eventually reach 0. Total population is hardly affected by the increase in incidence. It is worth noting in the second system, at the end of the period, the recovered group is still a significant portion of the total population.

This time, enter

$$\begin{array}{cccc} b = 0.525 & d = 0.5 & c = 0.001 & e = 0.5 \\ g = 0.1 & a = 0.2 & S_0 = 1000 & E_0 = 100 \\ I_0 = 50 & R_0 = 15 & A = 0.1 & T = 20 \end{array}. \quad (13.2)$$

and vary A using as the second choice $A = 2$. With a higher weight parameter, we can vaccinate at the maximum level for a longer period of time. This change greatly decreases the susceptible population and increases the recovered population. However, the exposed and infectious populations are reduced, but the change is marginal. Total population seems unaffected in any way, which seems to suggest early vaccinations are the key to disease management. Later vaccinations, while effective and helpful, become increasingly less efficient as time passes. To verify this, vary $A = 0.1$ versus $A = 200$. Here, with such a high A value, vaccination cost is of virtually no importance. As such, maximum vaccination is used almost exclusively. The effects on the susceptible and recovered populations are pronounced, but the change in the number of exposed and infectious people is small.

Enter (13.2) varying with $g = 0$, representing a disease where no recovery can occur. A higher vaccination rate must be used, as immunity is no longer achieved naturally. The second system has a higher infectious population throughout. This makes sense, as no one is recovering. The reduction in infectious people in the second system is due only to death. Now try (13.2) varying with $g = 0.4$. Here, one stands a much better chance of recovering from the disease. The infectious population reduces more rapidly, meaning a less aggressive vaccination routine can be used. Note, even though natural recovery is occurring more often in the second system, there are fewer recovered people. The shift in vaccination outweighs the shift in recovery rate.

We might suspect a higher disease-related mortality would necessitate a greater immunization rate. However, the opposite is actually true. Enter (13.2), this time varying with $a = 0.4$. In the second system, a slightly less powerful, but still aggressive initial immunization strategy is used. The infectious population reduces more rapidly due to the greater mortality rate, and fewer vaccinations are needed. Notice, however, that the total population in the second system is lowered slightly. If you try a disease mortality rate as high as $a = 1.5$, you will see very little vaccination is used, as the infectious population rapidly declines. The effect on total population also becomes more severe. Recall our objective functional minimizes the number of infectious people only. This simulation suggests we might also consider the total population in our goals.

Vary the latency of the disease, using (13.2) and varying with $e = 0.1$. In the second system, the disease has an incubation period five times as long. So, a large initial round of immunizations is not needed. The longer incubation period allows the immunizations to be spread out over the first several years. Also, the infectious population receives no initial boost and reduces at a greater speed. The recovery and death rates of infectious individuals are now larger than the rate at which susceptible people become exposed, then infectious.

Consider the relationship between the management of the disease and the effective growth of the population as a whole. In all the previous simulations, we considered a population with moderate growth. We now turn our attention to a simulation with rapid growth. Enter (13.2), varying with $b = 0.55$. Here, we have doubled the effective growth rate $(b - d)$. With so many more susceptible people, the disease can spread more easily. Thus, a more stringent schedule of immunizations must be used to balance out the population growth. Infectious and exposed populations are similar in both systems for the majority of the time interval. However, as immunization is decreased, both begin to rise at approximately 15 years. At this point, there are so many susceptible people, even a few infectious individuals are enough to restart the epidemic if immunizations are not continued. Conversely, consider a population with small or no effective growth, i.e., $b = d$. Enter the same (13.2) values as before, this time varying d to $d = 0.525$. The initial immunization blitz reduces the number of susceptible people as normal, but as the growth rate is zero, the susceptible population will have a slower increase. Thus, fewer vaccinations are needed after the first few years. The exposed and infectious populations are reduced in the usual way. However, the total population, without disease, is naturally static as $b = d$. Thus, the disease-caused deaths cause the total population to reduce in size.

To this point, we have examined populations where susceptible people were in the majority. Now, let us consider a case where the infection has been spreading unchecked for some time before intervention occurs. Enter the values

$$\begin{array}{|llll|} \hline b = 0.525 & d = 0.5 & c = 0.001 & e = 0.5 \\ g = 0.1 & a = 0.2 & S_0 = 1000 & E_0 = 1000 \\ I_0 = 2000 & R_0 = 500 & A = 0.1 & T = 20 \\ \hline \end{array}. \qquad (13.3)$$

Here, maximum vaccination is almost the entire period. The number of exposed and infectious people are reduced by the end of the period, but not nearly to the levels we have been observing. The number of exposed people actually begins to increase again in the last year. Most troublesome, the total population drastically falls in the first five years, before stabilizing and then increasing. Now try

$$\begin{array}{llll} b = 0.525 & d = 0.5 & c = 0.001 & e = 0.5 \\ g = 0.1 & a = 0.2 & S_0 = 1000 & E_0 = 2000 \\ I_0 = 5000 & R_0 = 1000 & A = 0.1 & T = 20 \end{array} \quad (13.4)$$

Here, vaccination has begun too late. Even with maximum vaccination for more than 19 of the 20 years, total population steadily falls. This again establishes the importance of early vaccinations. Treatment must begin before the infection gets out-of-hand. Try to create a set of parameters where the population has moderate growth but eventually dies out, despite the immunization tactics.

On your own, examine a few special cases of the initial conditions. Run a simulation where immunization begins before the disease becomes infectious, namely $I_0 = R_0 = 0$. Consider a closed environment, such as a cruise ship, where a few infectious individuals enter an uncontaminated population, specifically, $E_0 = R_0 = 0$. Also, try $E_0 = I_0 = R_0 = 0$.

Vary each of the initial conditions one by one to see their effect on the optimal immunization treatment. How does shortening the time interval alter the execution and efficiency of the immunization schedule?

Exercise 13.1 Consider this model with an objective functional that instead maximizes N. For example,

$$\max_u \int_0^T AN(t) - u(t)^2 \, dt$$

subject to
$S'(t) = bN(t) - dS(t) - cS(t)I(t) - u(t)S(t), \ S(0) = S_0 \geq 0,$
$E'(t) = cS(t)I(t) - (e+d)E(t), \ E(0) = E_0 \geq 0,$
$I'(t) = eE(t) - (g+a+d)I(t), \ I(0) = I_0 \geq 0,$
$R'(t) = gI(t) - dR(i) + u(t)S(t), \ R(0) = R_0 \geq 0,$
$N'(t) = (b-d)N(t) - aI(t), \ N(0) = N_0,$
$0 \leq u(t) \leq 0.9.$

Write a code for this problem (or alter code7.m), and examine the differences between the two problems.

Chapter 14

Lab 8: HIV Treatment

In the following lab, optimal control is used to find an optimal chemotherapy strategy in the treatment of the human immunodeficiency virus (HIV). Unlike the last lab, where the dynamics of a population affected by an epidemic were considered, this problem studies the immune system of an individual.

A great deal of research has been conducted on the effect of chemotherapy on the HIV virus. For example, in [105], Kirschner et. al. study the effects of chemotherapy on reducing viral production, which is most applicable to drugs such as protease inhibitors. Here, we consider the chemotherapy of reverse transcription inhibitors, such as AZT, which affects the "infectivity" of the virus. These drugs interrupt key stages of the infection process during the life cycle of HIV within a host cell. Butler, Kirschner, and Lenhart created a model for this type of interaction and used optimal control to develop treatment strategies in [25]. This lab is based on their work.

It is assumed the treatment acts to reduce the infectivity of the virus for a finite time, until drug resistance occurs. The measure of benefit of chemotherapy treatment is based solely on the increase of the $CD4^+T$ cell count. Thus, the model used describes the interaction of the immune system with HIV. Let $T(t)$ and $T_i(t)$ be the concentration of uninfected and infected $CD4^+T$ cells, respectively, and let $V(t)$ be the concentration of free virus particles. In this instance, concentration refers to the population count per unit volume. Let

$$\frac{s}{1+V(t)}$$

be the source term from the thymus, representing the rate of generation of new $CD4^+T$ cells. Let r be the growth rate of T cells per day. This growth is assumed to be logistic, with a maximum level of T_{max}. Let $kV(t)T(t)$ be the rate that free virus cells infect T cells. Let m_1, m_2, m_3 be the natural death rates of uninfected $CD4^+T$ cells (T), infected $CD4^+T$ cells (T_i), and free virus particles (V), respectively. Once infection of a T cell occurs, replication of the virus is initiated and an average of N virus particles are produced before the host cell dies.

The control, $u(t)$, is the strength of the chemotherapy, where $u(t) = 0$ is maximum therapy and $u(t) = 1$ is no therapy. We note that maximum therapy $u = 0$ is probably unrealistic to achieve; a more realistic positive lower bound would be better. We leave the problem as originally stated, though. A flow chart is given in Figure 14.1. We wish to maximize the number of uninfected

T cells while simultaneously minimizing the "cost" of the chemotherapy to the body. The fixed time frame simulates the period before drug resistance occurs. Letting $A \geq 0$ be the cost, or weight, parameter, the problem is

$$\max_u \int_0^{t_{final}} AT(t) - (1-u(t))^2 \, dt$$

subject to $T'(t) = \dfrac{s}{1+V(t)} - m_1 T(t) + rT(t)\left[1 - \dfrac{T(t)+T_i(t)}{T_{max}}\right]$
$\phantom{\text{subject to}\quad T'(t) =}- u(t)kV(t)T(t),$
$T_i'(t) = u(t)kV(t)T(t) - m_2 T_i(t),$
$V'(t) = Nm_2 T_i(t) - m_3 V(t),$
$T(0) = T_0 > 0,\ T_i(0) = T_{i0} > 0,\ V(0) = V_0 > 0,$
$0 \leq u(t) \leq 1.$

This example has only one control; see [90] for an HIV immunology problem with two controls representing two types of drug treatments. We note that models of HIV have changed since this work, but this still provides an excellent example of an immunology model.

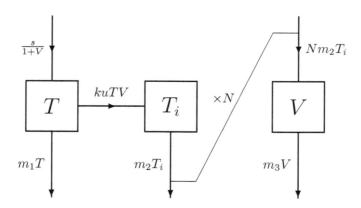

FIGURE 14.1: There are two types of T cells, uninfected and infected. Although the virus concentration is a separate population (T cells do not become virus particles), the death of infected T-cells directly affects the production of new virus particles. This is illustrated with the line from the death of infected T-cells to new virus particles, labeled with multiplication by N.

Lab 8: HIV Treatment

Enter MATLAB and begin *lab8*. Begin with the values

$$\begin{array}{cccc} s = 10 & m_1 = 0.02 & m_2 = 0.5 & m_3 = 4.4 & r = 0.03 \\ T_{max} = 1500 & k = 0.000024 & N = 300 & T_0 = 800 \\ T_{i0} = 0.04 & V_0 = 1.5 & A = 0.05 & t_{final} = 20 \end{array} \quad (14.1)$$

We see the optimal chemotherapy treatment begins with the strongest dose, followed by a decreasing of treatment. (Remember $u = 0$ is the maximum therapy, and $u = 1$ is no therapy). This has the effect of steadily increasing the T cell concentration, even though treatment is not 100% effective 100% of the time. This behavior is seen in drugs such as AZT and DDT. Also, infected T cell count and viral concentration initially decrease and then increase as treatment lessens, but only to a fraction of original levels.

Let us begin with an evaluation of the weight A. Enter the (14.1) values, varying with $A = 0.025$ as our second value. As expected, the system with higher cost parameter has a control where maximum treatment is continued longer. Subsequently, the T cell count is driven higher, and infected T cell and viral concentration are pushed lower. However, the increase in healthy T cells is marginal, while the infected T cell and viral concentrations are approximately halved. This is somewhat surprising, considering that only healthy T cell count is explicitly considered in the objective functional. Notice that both systems exhibit the same basic behavior. Optimal treatment for both begins with a period of maximum strength, then reduces in strength until reaching no treatment, both before the 20 day period is actually over.

Enter (14.1) again, this time varying the number of virus particles produced by infected cells, say $N = 250$ as the second value. Here, the concentrations of uninfected and infected T cells are approximately the same for both systems and the viral concentration is only slightly altered. However, the treatment regimen in the first system sustains maximum strength for a full day longer to achieve virtually the same results. Now, try $N = 300$ versus $N = 50$. Notice the dramatic difference. In the second system, the T cell count is driven a little higher, but with a much less strenuous treatment regimen. In fact, maximum strength treatment is never used, and treatment effectively ends after only ten days. Further, with $N = 50$, after the population is driven sufficiently low enough with drugs, the virus actually fades because it is not reproducing enough to sustain itself.

Now, enter (14.1), varying T cell growth rate to $r = 0.045$. Because of the higher natural growth rate of the T cells in second system, we are able to drive the T cell count higher, with virtually the same treatment regimen. It is worth pointing out the fundamental differences in this simulation and the last. When N was advantageously adjusted, the optimal treatment was to reduce drug strength in order to achieve the same T cell count. Conversely, when T cell growth rate was increased, the optimal action was to instead maintain a similar treatment schedule in order to gain a better T cell count. This is a direct result of the objective functional, which considers only healthy T cell count. An increased growth rate directly affects T, whereas a lower N value hinders viral production, altering T cells only indirectly.

Vary the infection rate, using (14.1) and varying with $k = 0.000032$. Increasing the infectivity of the virus has somewhat expected results. Viral and infected T concentrations are higher at the end of the time period, when drug treatment is greatly reduced. Also, uninfected T cell concentration remains approximately the same, while a longer period of maximum treatment is needed to achieve this. Again, notice the interesting duality, as, in this case, the optimal strategy is to increase overall drug strength in order to achieve the same healthy T cell count.

We now turn our attention to the generation of new $CD4^+T$ cells, represented by s. With (14.1), vary s to the second value of $s = 9$. Then, try $s = 8$ as the second value. Then $s = 7$. The behavior is unlike what we saw with N, k, and r. Here, as s varies, the optimal control and uninfected T cell count both change. A weaker drug regimen is used and the healthy T cell count decreases. When s is reduced, the production of T cells is slowed, so it is logical that the T cell count would be less in the second system. Notice the viral and infected T cell concentrations remain essentially unchanged through these variations of s. With fewer T cells, a weaker drug treatment is needed to keep the same levels of infection.

Try varying the maximum concentration of T cells in (14.1), using $T_{max} = 1200$ as the second value. One might suspect that since neither 1500 nor 1200 are particularly close to our initial condition of 800, this variation will cause little change in the outcome. However, you will see from the graphs that this is not true. The T cell count in the second system, which begins at 67% carrying capacity, actually decreases initially, before finishing at a level only slightly above the initial number. The first system T cell count, which begins at only 53% carrying capacity, increases steadily after the first two days. This occurs despite the two systems using very similar treatment regimens.

Now consider the length of the time interval. Vary the final time using (14.1) and $t_{final} = 10$. In the shortened time interval, the treatment makes far less progress. The final T cell count in the second system is lower than the first system on day 10. The treatment is even less effective in reducing the infected T cell and viral concentrations. Notice, however, that the maximum drug strength is held for a shorter interval and overall drug strength is less. The way the problem is cast, we are requiring drug side-effects to be minimized over a 10 day period. By comparing to a 20 day regimen, we are, in some sense, unfairly capping the allowed side-effects. The addition of 10 days to the treatment period will certainly lead to more side-effects. If, for instance, days 11 - 20 are completely drug-free, then we should allow twice the amount of side-effects in days 1 - 10 as compared to the original 10 day treatment. Hence, run the simulation with $t_{final} = 10$ and $A = 0.1$, twice the original value. Here, we see the dynamic we are used to. However, the final number, although better than the first 10 day schedule, are not comparable to the

20 day period. The length over which a drug is administered seems to be an integral part of its effectiveness. Thus, it is important to develop strong reverse transcription inhibitors which also have long resistance times.

The effect of varying each of the death rates m_1, m_2, and m_3 is predictable. Also, we have already examined the effect of initial conditions on systems such as this. So, the study of these parameters is left as an exercise for the reader. It may be of interest to ascertain which death rate and initial condition have the most impact.

Chapter 15

Lab 9: Bear Populations

This lab focuses on a metapopulation harvesting model, which will involve two separate controls. It is based on work done by Salinas, Lenhart, and Gross in [165]. A metapopulation is a population consisting of multiple local populations. Here, we wish to model a black bear population which is motivated by the scenario in the Great Smoky Mountain National Park. We consider the bear population in a "generic" park, in the surrounding national forest, and in the "outside" or "human-populated" areas (e.g., small towns and tourist areas). These areas have shared borders, and the density-dependent emigration and immigration of the bears is based on the connectivity of the patches.

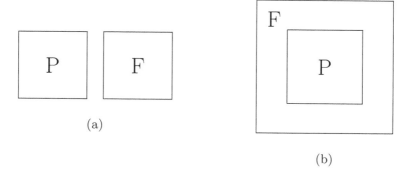

FIGURE 15.1: Park and forest positions for (a) $m_p = 0 = m_f$ (b) $m_p = 1$, $m_f = 0.35$.

Let r be the population growth rate of the bears, which will be assumed to be the same in the park and the forest. Let K be the carrying capacity. Let m_p be the proportion of the park boundary connected with the forest, and let m_f be the proportion of the forest boundary connected with the park. All boundaries not accounted for by m_p, m_f are assumed to connect to outside areas. Various border alignments are shown in Figures 15.1 and 15.2. Let $P(t)$, $F(t)$, and $O(t)$ be the bear density in the park, forest, and outside regions, respectively, at time t. We also consider the rates of bear harvesting in the park ($u_p(t)$) and in the forest ($u_f(t)$). Harvesting occurs via bear hunting in the forest, where hunting limits are enforced, and possible culling by the Forestry Service in the forest and park. Hunting is currently illegal in this national park, but we consider a model with harvesting allowed in the park. We wish to minimize the numbers of bears that enter the human-populated areas and thereby reduce bear-human encounters. To do this, we find the optimal harvesting levels in both the park and forest that minimize the outside population O, while taking the cost of harvesting into account,

$$\min_{u_p, u_f} \int_0^T O(t) + c_p u_p(t)^2 + c_f u_f(t)^2 \, dt$$

subject to $\quad P'(t) = rP(t) - \frac{r}{K}P(t)^2 + \frac{m_f r}{K}\left(1 - \frac{P(t)}{K}\right)F(t)^2 - u_p(t)P(t),$

$$F'(t) = rF(t) - \frac{r}{K}F(t)^2 + \frac{m_p r}{K}\left(1 - \frac{F(t)}{K}\right)P(t)^2 - u_f(t)F(t),$$

$$O'(t) = r(1 - m_p)\frac{P(t)^2}{K} + r(1 - m_f)\frac{F(t)^2}{K} + \frac{m_f r}{K^2}P(t)F(t)^2$$
$$+ \frac{m_p r}{K^2}P(t)^2 F(t),$$

$P(0) = P_0 \geq 0, \, F(0) = F_0 \geq 0, \, O(0) = O_0 \geq 0,$
$0 \leq u_p(t) \leq 1, \, 0 \leq u_f(t) \leq 1.$

In the objective functional, c_p, c_f are positive weight parameters. In the P state equation, rP represents the natural exponential growth,

$$\frac{r}{K}P^2$$

is the density-dependent emigration,

$$\frac{m_f r}{K}(1 - \frac{P}{K})F^2$$

is the density-dependent immigration from the forest, and $u_p P$ is the level of harvesting. The terms are similar in the state equation for F. In the O state equation,

$$r(1 - m_p)\frac{P^2}{K}$$

represents emigration from the park to outside areas, while

$$r(1 - m_f)\frac{F^2}{K}$$

is emigration from the forest to outside areas. The terms

$$\frac{m_f r}{K^2} PF^2 \quad \text{and} \quad \frac{m_p r}{K^2} P^2 F$$

are, respectively, the portion of the emigrating population from the park that enters the forest but is forced to leave because of density-dependence, and the portion entering the park from the forest that is forced to leave.

Open MATLAB and enter *lab9*. First, consider a portion of the park which shares half its border with the forest, and vice versa. At this time, the Great Smoky Mountain National Park does not participate in the culling of black bears, either in the forest or in the park. It is unlikely they will change this policy except in case of extreme overpopulation or other crises. To simulate a situation like that, we impose a very large cost on park harvesting. Begin with the values

$$\boxed{\begin{array}{lllll} r = 0.1 & K = 0.75 & m_p = 0.5 & m_f = 0.5 & P_0 = 0.4 \\ F_0 = 0.2 & O_0 = 0 & c_p = 10000 & c_f = 10 & T = 25 \end{array}} \quad (15.1)$$

Here, the harvesting inside the park is effectively zero. The park population grows logistically towards the carrying capacity 0.75. The forest harvesting is concentrated early in the period, causing the bear population in the forest to be almost constant initially. Afterwards, however, the forest population begins to grow, which, along with the growth in the park, causes a rapid rise in the number of bears in the outside areas.

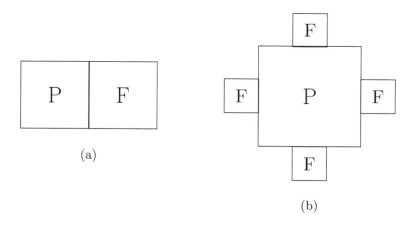

FIGURE 15.2: Park and forest possible positions for (a) $m_p = 0.25 = m_f$ (b) $m_p = 1/3$, $m_f = 0.25$.

Now enter the values (15.1) again, this time varying with $K = 2$. The large carrying capacity reduces the levels of immigration and emigration. The bear populations in the park and in the forest are increased over time, as fewer bears leave the area. Consequently, the outside population is reduced, while requiring less harvesting in the forest. Notice in both systems, the park population shows signs of leveling off, while the forest population is growing exponentially. This is due to the low initial concentration of the forest population, which is far from carrying capacity. Now vary with $K = 0.5$, a carrying capacity much closer to the initial concentrations. In the second system, the forest harvesting begins at a higher level, but decreases below the levels of the first system after about 10 years. Both the park and forest populations experience slower growth in the second system, as they are near carrying capacity. This causes an initial boost in the outside population, but by the end of the time period, the second system actually has fewer bears in the human areas.

This time, enter (15.1), varying with $r = 0.2$. The harvesting level in the park is increased, but still effectively zero. The higher growth rate causes an early population boom in the park, but the growth subsides as the carrying capacity is approached. This leads to greater emigration to the forest and outside areas. The higher growth rate and immigration into the forest necessitate a more active forest harvesting rate. Still, though, the forest concentration increases from the first system. Higher population concentrations lead to an increased number of bears leaving both the park and the forest for outside areas. By the end of the period, the number of bears in the outside area has tripled from the first to second system.

What if the park and forest bordered each other differently? Suppose the forest borders the park on all sides but one, $m_p = 0.75$ and $m_f = 0.3$ for example. Enter

$$\begin{array}{|cccccc|} \hline r = 0.1 & K = 0.75 & m_p = 0.75 & m_f = 0.3 & P_0 = 0.4 \\ F_0 = 0.2 & O_0 = 0 & c_p = 10000 & c_f = 10 & T = 25 \\ \hline \end{array} \quad (15.2)$$

You will notice the graphs are virtually the same. You will also see they remain essentially unchanged if you consider a situation where the park is entirely surrounded by the forest, such as

$$\begin{array}{|cccccc|} \hline r = 0.1 & K = 0.75 & m_p = 1 & m_f = 0.4 & P_0 = 0.4 \\ F_0 = 0.2 & O_0 = 0 & c_p = 10000 & c_f = 10 & T = 25 \\ \hline \end{array} \quad (15.3)$$

You might assume from this the border parameters have little effect on the system. However, as we will see, the underlying effect can be verified by varying other parameters. Reenter (15.3) and vary with $c_p = 100$. In the second system, we are "allowing" harvesting in the park. As expected, the harvesting level greatly increases, which reduces the bear population in the park over the entire time interval. As there are less bears to emigrate from the park, the forest and outside populations are reduced, and less harvesting in the forest is needed. Now suppose the park and forest share only a tiny portion of their borders. Enter

$$\boxed{\begin{array}{llllll} r = 0.1 & K = 0.75 & m_p = 0.1 & m_f = 0.1 & P_0 = 0.4 \\ F_0 = 0.2 & O_0 = 0 & c_p = 10000 & c_f = 10 & T = 25 \end{array}} \quad (15.4)$$

and vary c_p as before, with $c_p = 100$. The park and outside population, along with the level of park harvesting, are affected exactly as before. This time, however, the forest population and harvesting levels experience very little change. The thinning of bears in the park has little effect on the forest if the border between them is too small.

Enter the (15.1) values again, varying with $T = 15$. Both park harvesting rates are effectively zero. Notice the park populations are almost identical over the common time interval. As there is essentially no park harvesting, the bear population in the park grows near the same rate for both systems. There is more change in the outside population, which is caused by greater immigration from the forest. The shorter time interval forces the forest harvesting level to decrease, in turn increasing the forest bear population. In fact, the final forest bear population in the second system is almost equal to the final forest bear population in the first system, ten years sooner.

Finally, we examine the role of the initial conditions. Enter (15.1), first varying with $P_0 = 0.5$. The higher initial park population results in the second system having more park bears throughout, though the populations become closer in size as both approach the carrying capacity. The forest population and harvesting levels are slightly affected by the greater rates of immigration from the park, and the outside population is pushed up accordingly. Now vary the initial forest population, with $F_0 = 0.3$. The result is quite different than before. Here, more aggressive harvesting is done early to negate the effect of the higher starting population. By year 15, the forest population and harvesting in both systems are nearly identical. The park and outside populations, in addition to park harvesting, are only slightly affected. Finally, vary the initial outside population, using $O_0 = 0.1$. Notice only the outside population is affected; all other quantities are exactly the same in both systems. This behavior could have been predicted by examining the state equations. Notice P, F affect O, but O does not affect P or F. Also, notice the growth rate of O does not depend on its own size. We see in the graph that the outside populations in the two systems remain the same distance apart over the entire time interval. Their growth rate is the same throughout, even though the actual population sizes are not.

Chapter 16

Lab 10: Glucose Model

We mentioned earlier that systems of differential equations modeling specific behavior are often times very sensitive to changes in parameters. By now, you may have encountered solutions which were unrealistic or had problems which failed to converge when you supplied your own parameter values. Ideally, mathematical models are calibrated using data from field or clinical research. A model's effectiveness is based on its ability to accurately portray behavior inside the realm of the original data. Providing parameters which are well beyond these bounds can cause the system to act unexpectedly. In optimal control problems, the optimality system can yield an optimal control which makes little sense physically; sometimes, the system fails to converge to provide any solution at all. In this lab, we examine an ill-conditioned problem with this type of behavior.

In a study by Ackerman et al., a simplified, but highly accurate model of the blood regulatory system was developed to improve the ability of the GTT (glucose tolerance test) to detect pre-diabetics and mild diabetics [1]. The model considers the concentration of blood glucose g and net hormonal concentration h. It was shown that

$$g'(t) = c_1 g(t) + c_2 h(t),$$
$$h'(t) = c_3 g(t) + c_4 h(t).$$

Using properties of the blood glucose regulatory system, it is determined that $c_1, c_2, c_3 < 0$, and $c_4 \geq 0$. Also, it was shown that for diabetics, $c_4 = 0$ is a reasonable assumption.

This model is used by Eisen in an attempt to better regulate blood glucose levels in diabetic patients [55]. He introduces the change of variables $g = x_1$ and $h = x_2$, a convention we follow here. The goal is to find the insulin injection level, $u(t)$, which will minimize the difference between x_1 and the desired constant glucose level l, while taking "cost" of the treatment into account. Thus, the problem becomes

$$\min_u \int_0^T A(x_1(t) - l)^2 + u(t)^2 \, dt$$

subject to $x_1'(t) = -ax_1(t) - bx_2(t), \ x_1(0) = x_{10} > 0,$
$x_2'(t) = -cx_2(t) + u(t), \ x_2(0) = 0,$
$a, b, c > 0, \ A \geq 0.$

If you were to write out the optimality system for this problem, you would see that it is quite simple. In fact, it is a linear system with constant coefficients. If a, b, and c were specified, you could easily solve it by hand, finding the eigenvalues of a 4×4 matrix. As always, though, a solver code is provided. Enter MATLAB and begin *lab10*. Enter the following values

$$\boxed{a = 1 \quad b = 1 \quad c = 1 \quad x_{10} = 0.75 \quad A = 2 \quad l = 0.5 \quad T = 20}. \quad (16.1)$$

For now, do not vary any parameters. The results are somewhat troubling. The hormonal concentration and insulin level are both negative on the interval. This is clearly a physical impossibility. Also, the behavior of the glucose level seems to defy our goal. It decreases to the desired 0.5 level, then continues to decrease past it.

One problem could be the lack of control restrictions. Why not just require $u(t) \geq 0$? If the statement of the problem is so changed, and the code is appropriately adjusted, then it will simply converge to $u \equiv 0$. We might have been able to guess this would not work from the graph, where u is nowhere positive. The source of the problem is actually the parameters we have used, mainly the length of time. As we said earlier, the ODE model used in this optimal control problem was created for more accurate GTTs. While the model has proven very accurate for this purpose, in this problem we have asked much more. A GTT requires the monitoring of glucose for only three to five hours. Our problem requires the accurate prediction of glucose levels for weeks or even months. We have pushed the model well beyond its boundaries of reliability.

Now, enter the parameter values

$$\boxed{a = 1 \quad b = 1 \quad c = 1 \quad x_{10} = 0.75 \quad A = 2 \quad l = 0.5 \quad T = 0.2}. \quad (16.2)$$

Do not try to simultaneously view the solutions for $T = 20$ and $T = 0.2$, or you will not actually be able to see the second system. Of course, 0.2 days is 4.8 hours, which is within the acceptable time range for the model. As such, the results are very reasonable. The glucose concentration is steadily decreased, almost linearly, and the hormonal concentration and insulin levels are positive. Also, the scales of the hormone and insulin levels are more accurate. So, the model behaves perfectly and provides accurate solutions when reasonable parameters are used, but can give physically implausible solutions otherwise. Results for various T values are shown in Figure 16.1.

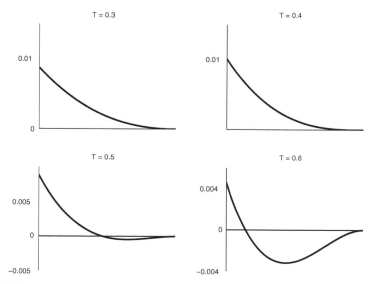

FIGURE 16.1: The optimal control u^* for four different T values, with $a = b = c = 1$, $x_0 = 0.75$, $A = 2$, and $l = 0.5$. We see the control becomes negative when $T = 0.5$, but it is still feasible for $T = 0.3$ and $T = 0.4$.

The purpose of this lab was to illustrate the necessity of reasonable parameters. We refrain from doing so here, but the reader may still find it insightful to experiment with this model, varying the parameters, as we have been doing.

Exercise 16.1 As mentioned earlier, this optimal control problem can be solved by hand. Find the optimal control in the case of the values (16.1). In particular, solve

$$\min_{u} \int_0^{20} 2(x_1(t) - 0.5)^2 + u(t)^2 \, dt$$

$$\text{subject to} \quad x_1'(t) = -x_1(t) - x_2(t), \, x_1(0) = 0.75,$$
$$x_2'(t) = -x_2(t) + u(t), \, x_2(0) = 0.$$

Chapter 17

Linear Dependence on the Control

In the preceding chapters, we have examined increasingly more general optimal control problems. However, we now turn our attention to a special case, which often arises in applications. Specifically, we focus on problems that are linear in the control u. The method of solving such problems is sometimes quite different, and the optimal solution often involves discontinuities in u^*.

17.1 Bang-Bang Controls

Consider the optimal control problem

$$\max_u \int_{t_0}^{t_1} f_1(t,x) + u(t)f_2(t,x)\,dt$$

$$\text{subject to} \quad x'(t) = g_1(t,x) + u(t)g_2(t,x), \ x(0) = x_0,$$

$$a \le u(t) \le b.$$

Notice the integrand function f and the right-hand side of the differential equation g are both linear functions of the variable u. Thus, the Hamiltonian is also a linear function of u, and can be written

$$H = [f_1(t,x) + \lambda(t)g_1(t,x)] + u(t)[f_2(t,x) + \lambda(t)g_2(t,x)].$$

The necessary condition $\lambda'(t) = -\frac{\partial H}{\partial x}$ is as normal. However, the optimality condition

$$\frac{\partial H}{\partial u} = f_2(t,x) + \lambda(t)g_2(t,x),$$

contains no information on the control. We must try to maximize the Hamiltonian H with respect to u using the sign of $\frac{\partial H}{\partial u}$, but, when $f_2 + \lambda g_2 = 0$, we cannot immediately find a characterization of u^*.

Define $\psi(t) = f_2(t, x(t)) + \lambda(t)g_2(t, x(t))$, usually calld the *switching function*. Our characterization of u^* is

$$u^*(t) = \begin{cases} a & \text{if } \psi(t) < 0 \\ ? & \text{if } \psi(t) = 0 \\ b & \text{if } \psi(t) > 0. \end{cases}$$

If $\psi = 0$ cannot be sustained over an interval of time, but occurs only at finitely many points, then the control u^* is referred to as *bang-bang*. In this case, it is piecewise constant function, switching between only the upper and lower bounds. An example of such a control is given in Figure 17.1. The switches coincide with the places where ψ switches signs (so that $\psi = 0$), hence the name switching function. The actual points where this occurs are called *switching times*. As mentioned in Sections 1.1 and 8.1, we will not be concerned with the actual value of the control at these times.

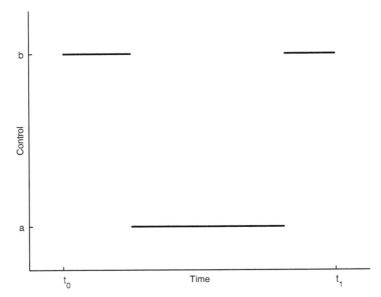

FIGURE 17.1: A typical bang-bang control.

If $\psi(t) \equiv 0$ on some interval of time, we say u^* is *singular* on that interval. A characterization of u^* on this interval must be found using other information. The endpoints of this interval are sometimes called switching times as well. We postpone the discussion of singular controls until the next section.

To solve a bang-bang problem numerically, the forward-backward sweep method can be employed. First, it must be analytically proven that the problem is in fact bang-bang, i.e., $\psi \equiv 0$ over an interval is impossible. Once this is established, the code is written as usual, where the characterization of

u is given by

```
for i=1:N+1
    temp = psi(t(i),x(i),lambda(i))
    if(temp < 0)
        u1(i) = a;
    else
        u1(i) = b;
    end
end
u = 0.5*(u1 + oldu);
```

where *psi(t(i),x(i),lambda(i))* in the second line is replaced by the actual value of the function ψ in terms of t, x, and λ, according to the specific problem. Notice, even though it is irrelevant from an analytical standpoint, the value of the control at the switching times must be assigned in our MATLAB code. Here, we have arbitrarily assigned $u = b$ when $\psi = 0$. Assigning $u = a$ would have been just as prudent. Defining u to be the average of a and b at these points is also used. It usually makes little difference. Finally, note the convex combination remains as before. This hastens the finding of the switching times.

Labs 11 and 12 investigate problems which are linear in the control. Both of these problems have bang-bang optimal controls. This is verified analytically in the lab before we examine the problems numerically. Here, we consider a few bang-bang examples which can be solved by hand.

Example 17.1

$$\max_u \int_0^2 e^t(1 - u(t))\, dt$$

$$\text{subject to } x'(t) = u(t)x(t),\ x(0) = 1,$$
$$0 \le u(t) \le 1.$$

The objective functional here does not depend on x, and the state does not have a terminal time condition. Therefore, looking at the format of the integrand of the objective functional we see that the optimal control should be 0.

The Hamiltonian is

$$H = e^t(1 - u) + \lambda u x.$$

The adjoint and transeversality conditions are

$$\lambda' = -\frac{\partial H}{\partial x} = -\lambda u, \quad \lambda(2) = 0,$$

and

$$\psi(t) = \frac{\partial H}{\partial u} = -e^t + \lambda(t)x(t).$$

Suppose u^* is singular on some interval, i.e., $0 < u^* < 1$. Then, $\psi = 0$ on this interval, so that

$$e^t = \lambda x.$$

As this holds on an interval, we can differentiate both sides,

$$e^t = (\lambda x)' = \lambda' x + \lambda x' = -\lambda u x + \lambda u x = 0.$$

This is clearly impossible, so u^* is nowhere singular, thus bang-bang. Considering both possible values for u^*,

$$u^* = 0 \Rightarrow x' = 0 = \lambda' \Rightarrow x, \lambda \text{ constant},$$
$$u^* = 1 \Rightarrow \lambda' = -\lambda \Rightarrow \lambda(t) = ke^{-t},$$

for some constant k. Note, for $\lambda(t) = ke^{-t}$ to satisfy $\lambda(2) = 0$, k must be zero. In the other case, λ is constant. Hence, regardless of what the control is near $t = 2$, $\lambda \equiv 0$ on some interval including $t = 2$. However, we require λ to be continuous, and it is impossible for $\lambda = ke^{-t}$ to be continuously joined with $\lambda \equiv 0$ for non-zero k. Thus, $\lambda \equiv 0$ everywhere. It follows

$$\frac{\partial H}{\partial u} = -e^t < 0 \text{ for all } t,$$

so that

$$u^* \equiv 0 \quad \text{and} \quad x^* \equiv 1.$$

Example 17.2 (from [100])

$$\max_u \int_0^2 2x(t) - 3u(t)\, dt$$
$$\text{subject to} \quad x'(t) = x(t) + u(t),\ x(0) = 5,$$
$$0 \leq u(t) \leq 2.$$

If we view this as a simple population model with exponential growth, we seek to increase the population as much as possible, while keeping the cost of control down.

The Hamiltonian is

$$H = 2x - 3u + x\lambda + u\lambda.$$

Using the necessary conditions and transversality condition, λ can be immediately solved:

$$\lambda' = -\frac{\partial H}{\partial x} = -2 - \lambda, \quad \lambda(2) = 0 \quad \Rightarrow \quad \lambda(t) = 2e^{2-t} - 2.$$

The switching function

$$\psi(t) = \frac{\partial H}{\partial u} = \lambda - 3 = 2e^{2-t} - 5$$

is clearly nowhere constant, thus not identically 0 over an interval. So, u^* is bang-bang, and

$$u^*(t) = 0 \Leftrightarrow \psi < 0 \Leftrightarrow e^{2-t} < 5/2 \Leftrightarrow t > 2 - \ln(5/2),$$
$$u^*(t) = 2 \Leftrightarrow \psi > 0 \Leftrightarrow e^{2-t} > 5/2 \Leftrightarrow t < 2 - \ln(5/2).$$

For $0 \leq t < 2 - \ln(\frac{5}{2})$,

$$u = 2 \Rightarrow x' = x + 2.$$

Along with $x(0) = 5$, this gives $x(t) = -2 + 7e^t$. On $2 - \ln(\frac{5}{2}) < t \leq 2$,

$$u = 0 \Rightarrow x' = x \Rightarrow x(t) = k_0 e^t$$

for some constant k_0. As x must be continuous, the expressions $-2 + 7e^t$ and $k_0 e^t$ must agree at $t = 2 - \ln(\frac{5}{2})$. This gives $k_0 = 7 - 5e^{-2}$. Hence, the optimal solutions are

$$u^* = \begin{cases} 2 & \text{when } t < 2 - \ln(5/2), \\ 0 & \text{when } t > 2 - \ln(5/2), \end{cases} \quad \text{and}$$

$$x^* = \begin{cases} 7e^t - 2 & \text{when } t \leq 2 - \ln(5/2), \\ 7e^t - 5e^{t-2} & \text{when } t \geq 2 - \ln(5/2). \end{cases}$$

The optimal state is shown in Figure 17.2.

17.2 Singular Controls

We now turn our attention towards singular controls, and in particular, a few examples. In the first example, the solution is relatively easy to guess.

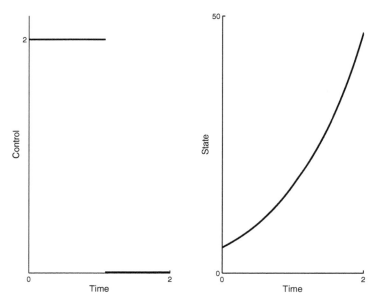

FIGURE 17.2: The optimal control and state for Example 17.2. The state appears differentiable here, but this is due to scale. It is in fact only continuous at $t = 2 - \ln(5/2)$.

However, because it is singular, generating the optimal control via the necessary conditions is somewhat difficult.

Example 17.3

$$\min_u \int_0^2 (x(t) - t^2)^2 \, dt$$

subject to $\quad x'(t) = u(t), \; x(0) = 1,$
$\qquad\qquad\quad 0 \leq u(t) \leq 4.$

First, generate the necessary conditions as usual,

$$H = (x - t^2)^2 + \lambda u,$$
$$\lambda'(t) = -\frac{\partial H}{\partial x} = -2(x - t^2), \quad \lambda(2) = 0,$$
$$\psi = \frac{\partial H}{\partial u} = \lambda.$$

If $\psi \equiv 0$ on some interval, then

$$0 \equiv \lambda'(t) = -2(x - t^2) \quad \Rightarrow \quad x(t) = t^2,$$

so that on this interval

$$u = x' = 2t.$$

Hence, we obtain

$$u^*(t) = \begin{cases} 0 & \text{when } \lambda > 0, \\ 2t & \text{when } \lambda = 0, \\ 4 & \text{when } \lambda < 0. \end{cases} \tag{17.1}$$

Our first goal is to establish that $x^*(t) \geq t^2$ on $[0, 2]$. Suppose not, i.e., suppose that $x(t) < t^2$ somewhere. Then, as $x(t) > t^2$ at $t = 0$, there must exist a $t_0 \in (0, 2)$ such that $x(t_0) \leq t_0^2$ and $u(t_0) = x'(t_0) < 2t_0$. Hence, from (17.1), it follows $u(t_0) = 0$ and $\lambda(t_0) > 0$.

Now, consider the points in time $t > t_0$ for which $\lambda(t) = 0$. We know at least one such point exits, namely $t = 2$. Let t_1 be the minimum of these points so that $\lambda(t_1) = 0$ but $\lambda(t) > 0$ for $t \in [t_0, t_1)$. Then, from (17.1), we see $u^* = 0$ on $[t_0, t_1)$. This implies $x^*(t) = x^*(t_0)$ on $[t_0, t_1)$. As we choose t_0 so that $x^*(t_0) \leq t_0^2$, we see $x^*(t) \leq t^2$ on $[t_0, t_1)$. Hence, by the adjoint equation, $\lambda'(t) \geq 0$ on $[t_0, t_1)$. But, if $\lambda(t_0) > 0$ and λ never decreases on this interval, then $\lambda(t_1) = 0$ is impossible. This gives our contradiction.

Thus, $x^*(t) \geq t^2$ on $[0, 2]$. This immediately gives $\lambda' \leq 0$ on $[0, 2]$. As $\lambda(2) = 0$, we must have $\lambda \geq 0$ on $[0, 2]$. As λ is a non-negative, non-increasing function, there is some $k \in [0, 2]$ so that $\lambda > 0$ on $[0, k)$ and $\lambda = 0$ on $[k, 2]$.

Suppose $k = 0$. Then, $\lambda = 0$ everywhere, so that $\lambda' = 0$ everywhere. But, $x^*(t) > t^2$ at $t = 0$ so that $\lambda'(0) < 0$. Contradiction. Now suppose $k = 2$. Then, $u^* = 0$ everywhere, so that $x^* = 1$ everywhere. This clearly contradicts $x^* \geq t^2$. Hence, $0 < k < 2$, and we have

$$u^*(t) = \begin{cases} 0 & \text{when } 0 \leq t < k, \\ 2t & \text{when } k < t \leq 2, \end{cases} \quad x^*(t) = \begin{cases} 1 & \text{when } 0 \leq t \leq k, \\ t^2 + (1 - k^2) & \text{when } k \leq t \leq 2, \end{cases}$$

as x^* must be continuous. Finally, to find k, note that $\lambda \equiv 0$ on $[k, 2]$, which implies

$$0 = \lambda'(t) = -2(1 - k^2) \text{ on } (k, 2) \quad \Rightarrow \quad k = 1.$$

Hence, the optimal solution set (Figure 17.3) is

$$u^*(t) = \begin{cases} 0 & \text{when } 0 \leq t < 1, \\ 2t & \text{when } 1 < t \leq 2, \end{cases} \quad x^*(t) = \begin{cases} 1 & \text{when } 0 \leq t \leq 1, \\ t^2 & \text{when } 1 \leq t \leq 2. \end{cases}$$

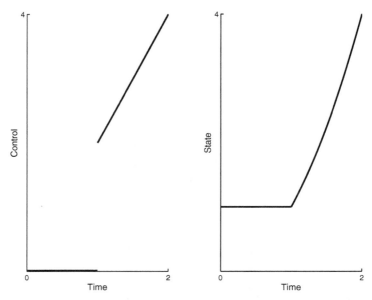

FIGURE 17.3: The optimal control and state for Example 17.3. Here, it is clear the state is not differentiable at $t = 1$.

The next example is a basic resource model example from Clark [33] and is more complicated than the previous example.

Example 17.4 (from [33])

$$\max_u \int_0^T \bigl(pqx(t) - c\bigr)u(t)\,dt$$

subject to $x'(t) = x(t)(1 - x(t)) - qu(t)x(t),\ x(0) = x_0 > 0,$
$0 \le u(t) \le M.$

We can think of this as a fishery model, where x represents a unit of harvested fish. Then, p is the price of one unit, q is the "catchability" of the fish, and c is cost of harvesting one unit. The control u is the effort put into harvesting. The integral represents total profit, revenue less cost. We have

$$H = (pqx - c)u + \lambda x - \lambda x^2 - qux\lambda,$$

$$\lambda' = -\frac{\partial H}{\partial x} = -pqu - \lambda + 2\lambda x + qu\lambda, \tag{17.2}$$

$$\psi = \frac{\partial H}{\partial u} = pqx - c - qx\lambda.$$

Consider the singular case, i.e., suppose $\psi \equiv 0$ on some interval. Assuming $c > 0$, this means $x \neq 0$. So, solving for λ we find

$$\lambda = \frac{pqx - c}{qx} = p - \frac{c}{qx}. \tag{17.3}$$

Differentiating this expression, and using the state equation for x', it follows

$$\lambda' = \frac{c}{qx^2}x' = \frac{c}{qx^2}(x - x^2 - qux) = \frac{c}{qx}(1 - x - qu). \tag{17.4}$$

By plugging the λ expression (17.3) into the adjoint equation (17.2), we get

$$\lambda' = -pqu - (p - \frac{c}{qx})(1 - 2x - qu). \tag{17.5}$$

Setting the expressions (17.4) and (17.5) equal to each other and doing some simple algebra, we find the u terms will cancel, and we arrive at the constant expression for the state below. Noting that $x' = 0$ during the singular interval, by plugging x^* into state equation, we can find u^*,

$$x^*(t) = \frac{c + pq}{2pq} \quad \text{and} \quad u^*(t) = \frac{pq - c}{2pq^2}.$$

First, note this problem may still be bang-bang. The singular value for the control is a constant. If this constant lies outside the bounds on the control, i.e., is less than 0 or greater than M, then the singular control is not achievable. This forces the optimal control to be bang-bang. Suppose, however, the singular control is possible, namely, that $0 < \frac{pq-c}{2pq^2} < M$. Then, we have a representation of u^* based on the value of the switching function as before. Can we solve for the optimal control numerically as we have been doing? In the bang-bang version of the forward-backward sweep, we assigned values to each point of the u^* vector depending on the sign of the switching function. Recall, we arbitrarily assigned the control to be at one of the bounds (or the average) when the switching function was zero. However, in singular problems, we have a separate characterization for u^* when the switching function is zero. Thus, it seems we could simply add a third value for u^*. For instance, in Example 17.4, our characterization would be

```
    for i=1:N+1
        temp = p*q*x(i) - c - q*x(i)*lambda(i)
        if(temp < 0)
            u1(i) = 0;
        elseif(temp == 0)
            u1(i) = (p*q - c)/(2*p*q^2);
        else
            u1(i) = M;
        end
    end
    u = 0.5*(u1 + oldu);
```

However, this method is an approximation, so it is unlikely the switching function ψ will ever be exactly zero. One possible solution to this problem is to use the singular value for u^* when ψ is in a small interval around zero, say $(-0.00001, 0.00001)$. If we tried this forward-backward routine on Example 17.4, we would find convergence occurs for all values of p, q, c, and M for which the control is bang-bang, and none of the values for which the control could be singular. Singular problems, because the switching function is identically zero on some interval, are often unstable for general control methods.

A great deal of work has been done on singular control problems, both analytically and numerically. Researchers have classified some of the behavior of singular optimal controls leading to additional necessary conditions, beyond those of Pontryagin's Maximum Principle, which singular optimal controls must satisfy. The most notable of these additional necessary conditions is the generalized Legendre-Clebsch condition [20, 29, 109, 116, 142]. This necessary condition is referred to as a *second order condition*, in that it involves higher order derivatives of the Hamiltonian. We have only used the second derivative of the Hamiltonian with respect to the control to verify that we are finding minimizing or minimizing controls. Second order optimality conditions, such as Legendre-Clebsch, use higher derivatives with respect to the states and time, not just the controls. These second order conditions have proven of great use in difficult singular problems, both analytically and numerically.

Many different types of numerical solvers have also been developed for problems which are linear in the control. Although we will not detail them here, they include gradient methods [156, 159], continuation methods [14, 31], iterative dynamic programming [132, 133], modified quasi-linearization methods [3], function space quasi-Newton methods [54], and adapted shooting methods [48, 49, 64, 140].

For some interesting results involving cancer chemotherapy and bang-bang and singular controls, see [44, 113, 114, 115]. Before concluding this chapter, we present an interesting example of a fishery problem in which x is the underlying variable.

Example 17.5 (from [154]) No-take marine reserves have been controversial, as advocates stress their conservation benefits and critics emphasize that the reserves decrease fishery yield. In a one-dimensional spatial problem, Neubert investigated the role of reserves as a part of an optimal harvest strategy designed to maximize yield [154].

After rescaling, the partial differential equation for stock density relative to the carrying capacity $w(x, t)$ is

$$w_t(x,t) = w_{xx}(x,t) + w(x,t)(1 - w(x,t)) - u(x)w(x,t).$$

Consider the corresponding steady state equation and call the stock density $w(x)$. Then,

$$0 = w_{xx}(x) + w(x)(1 - w(x)) - u(x)w(x)$$

with the boundary conditions

$$w(0) = w(l) = 0.$$

The boundary conditions reflect that the area surrounding the domain is uninhabitable. Note, this is a second order differential equation with the underlying variable x, and w_{xx} can instead be written as $w''(x)$. We assume the harvest control $u(x)$ satisfies $0 \leq u(x) \leq 1$.

Now converting to a system of two differential equations, we have

$$w'(x) = v(x), \; w(0) = w(l) = 0,$$
$$v'(x) = u(x)w(x) - w(x)(1 - w(x)).$$

We wish to maximize the yield over a fixed domain $[0, l]$ with controls satisfying $0 \leq u \leq 1$. So, our objective functional should be

$$J(u) = \int_0^l u(x)w(x)\,dx.$$

Note that for any control u, two functions which are identically zero give a solution to the state system. But here, we choose for our state variables the nonzero solutions of the state system (with w positive in the interior of the domain). We refer the reader to [26, 47] for the justification of such an approach.

The Hamiltonian and necessary conditions are easily found,

$$H = uw + \lambda_1 v + \lambda_2(uw - w(1-w)),$$
$$\lambda_1' = -\frac{\partial H}{\partial w} = \lambda_2(1 - 2w - u) - u,$$
$$\lambda_2' = -\frac{\partial H}{\partial v} = -\lambda_1, \ \lambda_2(0) = \lambda_2(l) = 0,$$
$$\psi = \frac{\partial H}{\partial u} = w(1 + \lambda_2).$$

If the switching function $w(1+\lambda_2)$ is zero on some interval and $0 < w < 1$ is valid on the whole interval, then the function $1 + \lambda_2$ must be zero. This means $\lambda_2 = -1$ and $\lambda_2' = 0$, which implies $\lambda_1 = 0$. Using the λ_1' equation, one obtains $w^* = \frac{1}{2}$ and then the state equation gives $u^* = \frac{1}{2}$ on this interval where the singular case holds. We conclude

$$u^*(t) = \begin{cases} 1 & \text{when } \lambda_2 > -1, \\ 1/2 & \text{when } \lambda_2 = -1, \\ 0 & \text{when } \lambda_2 < -1. \end{cases} \quad (17.6)$$

The necessary conditions need to solved numerically.

The results obtained in [154] depend on the length of the domain l, including some singular cases. From the numerical calculations in all cases, there is at least one interval where the optimal control is zero, which means a region of no-take reserve. For example, when the length of the domain is 6.3, there are 2 reserves, in the interior of the domain.

The case of no-flux boundary conditions, $v(0) = v(l) = 0$, is also quite interesting. The same argument as before still shows that $u = \frac{1}{2}$ is the singular case for an optimal control. But the objective functional can be used to obtain an exact solution in this case. By the state equations, $uw = v' + w(1-w)$, so that the objective functional can be rewritten

$$J(u) = \int_0^l v' + w(1-w)\,dx.$$

The v' term will vanish by integration by parts and the boundary conditions, and we are left with

$$J(u) = \int_0^l w(1-w)\,dx.$$

Irrespective of any optimal control techniques, it is clear this integral is maximized by $w^* = \frac{1}{2}$, because $w \geq 0$ everywhere. This gives $u^* = \frac{1}{2}$, and the singular case occurs on the whole time interval.

The management of fisheries is an area where economics comes naturally into the consideration. The classic book by Clark [33] is an excellent place to start looking at such models including economic impacts. See recent papers by Herrera and Sanchirico et. al. [83, 166] for results with economic effects and spatial considerations. See also [100, 167] for other economic applications.

17.3 Exercises

Exercise 17.1 (from [169]) Solve

$$\max_u \left[8x_1(18) + 4x_2(18) \right]$$

subject to $x_1'(t) = x_1(t) + x_2(t)$, $x_1(0) = 15$,
$x_2'(t) = 2x_1(t) - u(t)$, $x_2(0) = 20$,
$0 \leq u(t) \leq 1$.

Exercise 17.2 Solve Example 17.2 as a minimization problem.

Exercise 17.3 Solve

$$\min_u \int_0^1 u(t)\, dt$$

subject to $x'(t) = x(t) - u(t)$,
$x(0) = 1$, $x(1) = 0$,
$1 \leq u(t) \leq 2$.

Exercise 17.4 (from [100]) Solve

$$\min_u \int_0^1 (2 - 5t)u(t)\, dt$$

subject to $x'(t) = 2x(t) + 4te^{2t}u(t)$,
$x(0) = 0$, $x(1) = e^2$,
$-1 \leq u(t) \leq 1$.

Exercise 17.5 (from [175]) Solve

$$\min_u \int_0^2 x(t)^2\, dt$$

subject to $x'(t) = u(t),\ x(0) = 0,\ x(2) = 1,$
$\qquad\qquad -1 \le u(t) \le 1.$

Exercise 17.6 (from [175]) Let $r, s, p, b, x_0, T > 0$ be positive constants. Assume $\frac{b}{r+b} > 1 > s$ and solve

$$\max_u \int_0^T e^{-rt}(px(t) - u(t))\, dt + se^{-rT}x(T)$$

subject to $x'(t) = -bx(t) + u(t),\ x(0) = x_0,$
$\qquad\qquad 0 \le u(t) \le M.$

Exercise 17.7 (from [37]) Let a, b, x_{10}, x_{20} be constants where $a < b$. Solve

$$\min_u \int_0^5 4u(t) + x_1(t)\, dt$$

subject to $x_1'(t) = u(t) - x_2(t),\ x_1(0) = x_{10},$
$\qquad\qquad x_2'(t) = u(t),\ x_2(0) = x_{20},$
$\qquad\qquad a \le u(t) \le b.$

Chapter 18

Lab 11: Timber Harvesting

For our first lab involving a problem linear in the control, we examine a simple tree harvesting simulation. Consider a timber farm which, due to environmental regulations, can harvest at most a fixed percentage of its tree population, which must then be replanted. We assume the farm operates at this constant percentage of harvesting, producing raw timber in the amount of $x(t)$ at time t. The growth and death rates of individual trees are not considered. The amount of timber is based solely on the size of the farm (or equivalently, the number of trees). We also assume the harvest percentage level is low enough so that tree age need not be considered and there will always be mature trees ready for harvest. Once the timber has been processed, it is immediately sold. The money can either be kept as profit or reinvested in the farm by purchasing land and labor for further tree growth. The owners of the farm wish to find the reinvestment schedule which maximizes profit over a fixed time interval. Let the control $u(t)$ represent the percentage of timber revenue reinvested in the frame. Reinvestment will lead to the growth of more trees and the production of more timber, so we have $x'(t) = kx(t)u(t)$, where k is the return constant, which takes into account the average cost of labor and land. Further, if p is the market price of a unit of timber, then the profit at time t is $px(t)(1 - u(t))$, and the total profit is

$$p \int_0^T x(t)[1 - u(t)]\, dt.$$

However, we also wish to take into account money which could be gained through interest on profit. Profit earned in year one could be placed into an interest bearing account, while profit from the end of the time period could not. Therefore, money from earlier in the time frame is, in some sense, more valuable. In economics, this is referred to as *present-value*. If we let r be the interest rate over the period (assuming it is fixed), then the profit, adjusted for interest, is

$$p \int_0^T e^{-rt} x(t)[1 - u(t)]\, dt.$$

The exponential term in the integral is generally called a *discount term*. Exercise 17.6 in the previous chapter contained such a term. The function e^{-rt} is a decreasing function of time, encouraging money to be invested at the beginning of the interval. Future values of profit are discounted at a rate r. More information can be found in [100]. Finally, note that the constant p will not affect how this integral is maximized, so the optimal control problem is

$$\max_u \int_0^T e^{-rt} x(t)[1 - u(t)]\, dt$$

$$\text{subject to} \quad x'(t) = kx(t)u(t),\ x(0) = x_0 > 0,$$
$$0 \le u(t) \le 1.$$

The Hamiltonian is

$$H = e^{-rt} x(1-u) + kxu\lambda.$$

By developing the optimality system as usual, we find

$$\lambda' = u(e^{-rt} - k\lambda) - e^{-rt} \quad \text{and} \quad \psi = \frac{\partial H}{\partial u} = x(k\lambda - e^{-rt}).$$

As $x(0) = x_0 > 0$, $k > 0$, and $u \ge 0$ for all t, it follows that $x'(t) = kxu \ge 0$ and $x(t) > 0$ for all t. Suppose $\psi = 0$ over some interval. As x is strictly positive, this can occur if and only if $\lambda(t) = \frac{1}{k} e^{-rt}$ over some interval. Then, $\lambda'(t) = -\frac{r}{k} e^{-rt}$. However, if we use the adjoint equation, we instead find $\lambda'(t) = -e^{-rt}$. If $k \ne r$, this is clearly a contradiction. If $k = r$, then it follows that $\lambda(t) = \frac{1}{k} e^{-rt}$ for the remainder of the time interval. This contradicts $\lambda(T) = 0$. Thus, $\psi = 0$ cannot be sustained over an interval, and the optimal control is bang-bang.

To begin, type *lab11* at the prompt. Enter

$$\boxed{x_0 = 100 \quad k = 1 \quad r = 0 \quad T = 5}. \tag{18.1}$$

Do not vary any parameters for now. If you look at the control, you will notice that the graph is continuous and there is an abrupt shift at $t = 4$. Of course, we know the control should be bang-bang. MATLAB creates the graph by connecting a series of points, forcing it to be continuous. The part of the graph at $t = 4$, which is almost vertical, represents the switching point. Specifically, $u^*(t) = 1$ for $0 \le t < 4$ and $u^*(t) = 0$ for $4 < t \le 5$; $u^*(4)$ need not be defined. The state $x^*(t)$ is exponential for $0 \le t \le 4$ and constant for $4 \le t \le 5$, as expected. The solution tells us the optimal reinvestment strategy is to reinvest all timber revenue for the first four-fifths of the time interval, allowing the farm to grow. Then, near the end of the time interval, all revenue should instead be kept as profit.

Try the values

$$\boxed{x_0 = 100 \quad k = 0.3 \quad r = 0.05 \quad T = 5} \tag{18.2}$$

varying with $x_0 = 50$. Then, try $x_0 = 100$ vs. $x_0 = 500$. As the timber production grows exponentially, the two states are quite different. However,

Lab 11: Timber Harvesting

the two optimal controls are identical. Continue experimenting with the initial value. You will see the optimal reinvestment strategy is independent of initial timber production capacity.

Now, let us examine the role of k. Enter the (18.2) values, this time varying k with $k = 0.35$ as the second value. The higher k value represents a greater return with reinvestment, due to lower labor and/or land costs. We see that if we have a greater k value, a longer period of reinvestment and growth is optimal. The higher k allows the second state to grow more rapidly. At approximately $t = 1.355$, the first system ends reinvestment, but the second system continues for slightly longer. As the second state is already much higher at $t = 1.355$, the effect of continuing for this short time is much greater in system 2 than in system 1. This same type of behavior in the optimal control is also seen in the next lab. If you try $k = 0.3$ and $k = 0.6$, you will see the period of no investment at the end is almost cut in half. In fact, if we instead use $r = 0$, we will see doubling the value of k exactly cuts the interval of no investment in half; conversely, halving k doubles the size of the no investment interval. Finally, enter (18.2), varying with $k = 0.2$. Here, the return constant is so low, it is never advantageous to reinvest in the farm. All money should be kept as profit.

Enter (18.2), this time varying r with $r = 0.075$. This represents a greater return on the profits. As you might expect, the resulting optimal control is to end reinvestment in the farm earlier. With higher yield on profits, we wish to begin accumulating them earlier. Now vary with $r = 0.2$. Here, interest is so high it is best to take all money as profit.

Enter (18.2) varying the time interval with $T = 10$. Notice, the length of the no investment period in the controls is the same. This tells us the length of this period is a function of k and r only. Further, as we saw, with $r = 0$, it is inversely proportional to k. Now, try $T = 5$ versus $T = 3$. In the second system, the time interval is too short to conduct any reinvesting. The growth will take too long to make up for lost money. We could have predicted this, of course, based on what we saw above. In the first system, the length of the no investment period is approximately 3.645 years. As k and r alone determine this length, and neither was changed, the period of no investment in the second system must also be 3.645 years. However, the time interval is only 3 years. Thus, $u = 0$ must occur everywhere.

Finally, it is worth discussing briefly a numerical peculiarity that occurs with this problem. You may have already discovered that if you enter certain values convergence will not occur. One such set is $r = 0$ and $k = T = 1$ (x_0 is irrelevant). The reason for this failure is, surprisingly, round-off error. By observing the solutions when $k = 1$, $r = 0$, and $T \neq 1$ and then $T = 1$, $r = 0$, and $k \neq 1$, combined with our knowledge of how the solution behaves under changes in k and T when $r = 0$, we see the optimal solution with $k = T = 1$ and $r = 0$ is $u \equiv 0$. However, when the values of u are essentially 0, the adjoint equation becomes $\lambda'(t) = -1$. Using this and $\lambda(1) = 0$, the backward sweep of the code should arrive at $\lambda(t) = 1 - t$ and specifically $\lambda(0) = 1$.

However, it does not. After the backward Runge-Kutta sweep, the value of $\lambda(0)$ is actually $1 + 10^{-16}$. This may seem insignificant, but when used, gives $\frac{\partial H}{\partial u}(0) > 0$, causing $u(0)$ to be incorrectly stored as 1, not 0. Using this u, the subsequent sweeps will arrive at the correct control, namely $u \equiv 0$. However, the convex combination portion of the code will actually average the conflicting 0 and 1 values, causing the estimation of $u(0)$ to be 0.5, then 0.025, then 0.125, and so on. The value of $u(0)$ will slowly converge to the correct value of 0, but once it becomes small enough, essentially 0, the loop will begin anew, preventing convergence. This problem is actually caused entirely by the use of the Runge-Kutta 4 routine, specifically, the division by 6 in the algorithm. As $\frac{1}{6}$ is a repeating decimal, MATLAB cannot store the value with complete accuracy, leading to round-off error, which accumulates with each iteration. This problem also occurs when $r = 0$ and $T = 1/k$, for the same reasons. We should mention this type of convergence problem sometimes appears with bang-bang controls when a switching time occurs at the beginning or end of the interval, as it does here. Often times for bang-bang problems, a different type of code is used. Instead of forcing a representation of u^* to converge, an algorithm is used to estimate the switching times. More information on this method can be found in [52].

Exercise 18.1 The convergence problem discussed above will occur whenever the optimal control has a switching time at $t = 0$. The case $r = 0$ has already been explored. Find (by hand) the values of k and T which cause a switching time at 0 for $r > 0$ (x_0 is still irrelevant). Confirm using the code that convergence does not occur in this case.

Chapter 19

Lab 12: Bioreactor

This lab deals with a variation on Example 12.4. Often, contaminated soil will contain bacteria which, via metabolism or co-metabolism, is capable of eliminating the contaminant. Thus, a cost-effective method of managing and cleaning contaminated areas is to increase the level of these bacteria, in that more bacteria will result in more rapid degradation of the contaminant. The injection of nutrients needed for metabolism and colony growth has proven to be a successful technique in boosting the bacteria population. However, modeling the remediation process is nontrivial. Various processes, such as bacterial reproduction, metabolism, death, nutrient flow, and contaminant degradation, are highly coupled. Also of concern is the uncertainty in estimations of bacterial and contaminant distribution in the soil.

Hence, a realistic model would need to include spatial effects and heterogeneities in the environment. For this reason, we instead focus on the more controlled setting found in a bioreactor, such as the ones used for drug production and sewage treatment, for example. There, simple but effective relations for bacterial growth, degradation, and production processes can be formed. Precisely this type of model was developed and used for optimal control studies in the paper by Heinricher, Lenhart, and Solomon [82].

A bioreactor with ideal mixing is considered, where a contaminant and a bacteria known to degrade this contaminant via co-metabolism are present. The bacteria and contaminant have spatially uniform, time-varying concentrations (in g/L) $x(t)$, $z(t)$ respectively. The bioreactor is assumed to be rich in all nutrients needed for bacteria growth save one, whose injection into the reactor has spatially uniform, time-varying concentration $u(t)$. Bacterial growth rate is given by $Gu(t)x(t)$ and death rate by $Dx^2(t)$, so that

$$x'(t) = Gu(t)x(t) - Dx(t)^2, \quad x(0) = x_0 > 0.$$

The positive constant G represents maximum growth rate, while the positive constant D is the natural death rate. Note, as contaminant degradation occurs via co-metabolism, the contaminant is not needed for bacterial growth, and thus does not appear in the equation. The rate of degradation is assumed to be proportional to both bacterial and contaminant concentrations, in that

$$z'(t) = -Kx(t)z(t), \quad z(0) = z_0 > 0,$$

where $K > 0$ represents the degradation rate of the bacteria. The soil is assumed to have physical limitations that necessitate the bounds $0 \leq u(t) \leq M$ (the need to limit methane concentrations in the soil, for example). The goal is to minimize use of the nutrient and the final contaminant levels. As in Example 12.4, we consider the objective functional

$$\ln(z(T)) + \int_0^T Au(t)\,dt, \tag{19.1}$$

which should be minimized. Also as before, this allows a great simplification, in that we can eliminate z and its associated adjoint. Further, we set $A = 1$. The reason will be clear shortly. Solving $z'(t) = -Kx(t)z(t)$, we find

$$z(t) = z_0 \exp\left(-\int_0^t Kx(s)\,ds\right),$$

so that

$$\int_0^T Kx(s)\,ds = -\ln\left(\frac{z(T)}{z_0}\right) = -\ln(z(T)) + \ln(z_0).$$

Hence, our objective functional (19.1) is equal to

$$-\int_0^T [Kx(t) - u(t)]\,dt + \ln(z_0).$$

The addition of the constant $\ln(z_0)$ will have no effect on the minimization. Further, removing the negative sign and converting to maximization, an equivalent problem is

$$\max_u \int_0^T Kx(t) - u(t)\,dt$$

subject to $\quad x'(t) = Gu(t)x(t) - Dx^2(t),\ x(0) = x_0,$
$\qquad\qquad 0 \leq u(t) \leq M.$

It is now clear why A was set to 1; the variable K plays the role of a weight parameter here. Much like the maximum average mass in Lab 6, it plays two roles mathematically.

Before beginning, we must show the optimal control is bang-bang. The equation for $x'(t)$ is given, and the equation for the adjoint,

$$\lambda'(t) = -K - Gu\lambda + 2Dx\lambda,$$

is found as usual. Also,

$$\psi = \frac{\partial H}{\partial u} = -1 + G\lambda x.$$

Considering the function $(x\lambda)(t)$ and using the state and adjoint differential equations, we have

$$(x\lambda)' = x'\lambda + x\lambda' = Dx(x\lambda - K/D).$$

Now, it follows from the state equation and initial condition that $x(t) > 0$ for all t. So, by the differential equation, $x\lambda$ must be everywhere greater than, less than, or equal to the constant K/D. As $(x\lambda)(T) = 0 < K/D$, it follows $(x\lambda)(t) < K/D$ for all t. Thus, $x\lambda$ is monotonically decreasing, so that $A = G\lambda x$ could be maintained only at a single point. Hence, u^* is bang-bang, with the switching time occurring where $x\lambda - \frac{A}{G}$ changes sign. The variables u^*, x^*, and λ are as solved before. Once the optimal control, state, and adjoint are found, z^* is found using the differential equation.

Open MATLAB and enter *lab12*. Begin with the values

$$\boxed{K = 2 \quad G = 1 \quad D = 1 \quad x_0 = 0.5 \quad z_0 = 0.1 \quad M = 1 \quad T = 2}. \quad (19.2)$$

In this simulation, the optimal nutrient injection is no injection at all. The combination of the K and G values is such that the benefits of the nutrient does not outweigh its cost, so it is more efficient to leave the reactor undisturbed. This causes the bacterial population to steadily decrease. The bacteria present are able to degrade some contaminant, but the level of degradation decreases with the bacterial population. Now, enter the (19.2) values, this time varying K with a second value of $K = 2.1$. Only a slight change in K causes the injection of nutrients to become efficient. By increasing K, we have increased the importance placed on maximizing the level of bacteria, or, referring to the original problem, minimizing the final contaminant level. Here, it is advantageous to use nutrients to precipitate the growth of more bacteria. Also, whereas we might expect a small change in K to produce only a short period of injection, we see the period of nutrient addition is more than a quarter of the time interval. This is caused by the delayed reaction we see in contaminant concentration. The injection causes a sharp increase in bacteria, followed by the same steady decrease as in the other system. The contaminant level seems only slightly affected by the injection initially. However, the higher bacteria levels cause a more rapid contaminant degradation to build. By the end of the 2 day period, the second contaminant level is almost half that of the first. The optimal strategy is to inject the maximum nutrient levels early, which results in a sharp increase in bacteria population, which, in turn, triggers a quicker decrease in contaminant.

Note that in the graph, when the optimal control moves off its upperbound, it appears almost rounded as it decreases. We showed analytically the optimal control should be bang-bang. This rounding of the switching point is caused simply by numerical error. In large measure, it is due to the convex combination of the control estimates. The actual switching time is still approximately where the vertical line is. You will see this rounding effect in all simulations of this lab.

Enter (19.2) again, this time varying G to $G = 1.2$. We have increased the effectiveness of the nutrient in aiding the growth of new bacteria. As expected, this makes the injection of nutrients desirable, as the overall effect has been increased. Now enter the values

$$\boxed{K = 2.1 \quad G = 1 \quad D = 1 \quad x_0 = 0.5 \quad z_0 = 0.1 \quad M = 1 \quad T = 2} \quad (19.3)$$

varying G with $G = 1.2$. With this K, as we have seen, the injection of nutrients is optimal, even with $G = 1$. Thus, both systems use nutrient injection initially. However, during this period, the second system sees a greater increase in bacteria population, due to the greater growth rate G. Also, with the greater G, the injections continue for a longer period, as they have more effect.

Enter

$$\boxed{K = 2.1 \quad G = 1.2 \quad D = 1 \quad x_0 = 0.5 \quad z_0 = 0.1 \quad M = 1 \quad T = 2} \quad (19.4)$$

varying with $x_0 = 0.7$. We might expect a higher initial bacterial concentration to result in a shorter injection period. This way, less injection could be used to garner the same effect. However, we see the opposite occurs. The second system calls for a longer injection period. To understand this, we must remember the bacteria population grows and degrades the contaminant exponentially. So, at approximately 1.3 days, when injection ends in system 1, the effect of continuing injection in system 2 is greater than doing the same in system 1, as system 2 has more bacteria at this point. Enter (19.4) again, varying with $x_0 = 0.3$. Even with these K and G values, the second system calls for no injection. The low initial concentration means it would take much more nutrient supplementation to increase the bacteria to effective levels. So much more, apparently, that it is no longer efficient to do so.

Much like x_0, adjusting M does the opposite of what you may expect. Enter (19.4), varying with $M = 2$. One might guess that doubling the allowable injection strength would result in halving the injection period. Instead, though, the greater M results in a longer period of nutrient addition. This, again, is due to the exponential nature of the populations. The higher injection strength not only pushes the bacteria population higher, but to levels over three times that of the other system. Therefore, it is worth continuing injection for several hours longer. The final contaminant concentration in the first system is over 30 times the size of the final concentration in the second system.

Consider the effect of D on the optimal control. Enter the values (19.4), varying with $D = 0.8$. The lower death rate, in some sense, increases the effectiveness of the injection. Specifically, the same amount of injection will now result in a higher bacteria population. Like we have seen with x_0 and M, when injection becomes more effective, we should continue injections for a longer period.

Now try varying the initial contaminant concentration. Enter any two values for z_0. The optimal nutrient injections and resulting bacterial concentrations are identical. Only the contaminant concentrations are affected by the change in z_0. Considering the optimal control problem we have solved, this could have been predicted, as z_0 enters only the state equation for the contaminant level, and z was eliminated from the problem. However, if we instead recall the original problem, that of minimizing the final contaminant concentration and total cost of the nutrient injection, the irrelevancy of z_0 seems surprising. The value of z_0 directly affects the contaminant concentration. Nevertheless, we have shown the optimal strategy is unchanged by alterations of z_0. The actual contaminant levels are irrelevant. We should focus only on increasing the bacteria population.

Finally, enter (19.4) varying T with $T = 4$. There is little surprise here. The optimal strategy is to use the extra time to add more nutrients in order to build the bacteria population to greater levels. If you try other combinations of T values, the system with the longer time interval always has a smaller interval of no injection.

Exercise 19.1 Reconsider this problem without the introduction of the natural logarithm, i.e.,

$$\min_u z(T) + \int_0^T Au(t)\,dt$$

subject to $x'(t) = Gu(t)x(t) - Dx^2(t),\ x(0) = x_0,$
$z'(t) = -Kx(t)z(t),\ z(0) = z_0,$
$0 \le u(t) \le M.$

The same trick to eliminate the z variable can no longer be used. However, this problem can still be approached by the methods we have developed. Show that this optimal control problem is bang-bang, and write a MATLAB code to solve it. You will note that all the parameters have the same influence here as they did in the lab above, save one. In this problem, z_0 will affect the optimal control. Why do you think the introduction of the natural logarithm made z_0 irrelevant?

Chapter 20

Free Terminal Time Problems

In many applications, we are concerned with maximizing (or minimizing) an objective functional over a non-fixed time interval. If we return to our simple cancer example, Example 3.3, we could instead consider a slightly different problem. Before, we wanted to find a drug treatment over a given time frame $[0, T]$ which would minimize the final tumor cell concentration and total harmful effects of the drug. Suppose, instead, we want to find a time frame and a control that produce an objective functional value minimum among all time frames and all controls. Namely,

$$\min_{u,T} x(T) + \int_0^T u(t)^2 \, dt$$

subject to $x'(t) = \alpha x(t) - u(t)$, $x(0) = x_0$.

Notice that the minimization is now considered over the variables u and T. This is the standard way of writing an optimal control problem when T is free.

We now have more unknowns, with the optimal control and optimal terminal time both to be determined. To handle this problem, and other problems where the terminal time is free, we must redevelop the necessary conditions. As you will see, having given up information, in some sense, by allowing T to be free, we will gain new information in the way of a necessary condition we did not have before.

We note that we could just as easily allow the initial time, or both the initial and terminal times, to be free. In most applications, though, it is the final time which is allowed to move, so we handle this case.

20.1 Necessary Conditions

Let $f(t, x, u)$ and $g(t, x, u)$ be continuously differentiable functions in all three variables, and consider the free terminal time problem

$$\max_{u,T} \int_{t_0}^T f(t, x(t), u(t)) \, dt + \phi(T, x(T))$$

subject to $x'(t) = g(t, x(t), u(t))$, $x(t_0) = x_0$.

As there are two unknowns here, we write the value of the objective functional as

$$J(u, T) = \int_{t_0}^{T} f(t, x(t), u(t))\, dt + \phi(T, x(T)),$$

where, of course, x is the state corresponding to u. Let (u^*, T^*) be an optimal pair. Namely, u^* is a control on the nonempty, finite interval $[t_0, T^*]$ and $J(u, T) \leq J(u^*, T^*) < \infty$ for all other controls u and times T. Let x^* be the corresponding state. Let h be a piecewise continuous function and ϵ a real number. Then, $u^\epsilon(t) = u^*(t) + \epsilon h(t)$ is a control. As $J(u, T)$ reaches a maximum at u^*, T^*, we have that

$$0 = \lim_{\epsilon \to 0} \frac{J(u^*, T^*) - J(u^\epsilon, T^*)}{\epsilon}.$$

It follows from the same arguments used in Chapters 1 and 3 that

$$0 = \frac{\partial H}{\partial u} \text{ at } u^*,$$
$$\lambda' = -\frac{\partial H}{\partial x} = -f_x - \lambda g_x,$$
$$\lambda(T^*) = \phi_x(T^*, x(T^*)),$$

where ϕ_x refers to the partial derivative of ϕ in the state variable or the second variable. However, this still does not give any information about the optimal final time T^*. For this, we exploit the T variable of J. Consider real numbers $\delta \geq t_0 - T^*$, so that $T^* + \delta$ is an admissible terminal time. It is necessary to consider u^* and x^* on an interval larger than $[t_0, T^*]$. First, we can assume that u^* is left-continuous at T^*, by simply reassigning its value there if necessary. Then, set $u^*(t) = u^*(T^*)$ for all $t > T^*$, so that u^* will be continuous at T^*. Now, x^* is also defined for $t > T^*$. As $J(u, T)$ reaches its maximum at u^*, T^*, we have

$$0 = \lim_{\delta \to 0} \frac{J(u^*, T^* + \delta) - J(u^*, T^*)}{\delta},$$

or equivalently,

$$0 = \lim_{\delta \to 0} \frac{1}{\delta} \left[\int_{t_0}^{T^*+\delta} f(t, x^*, u^*) \, dt + \phi(T^* + \delta, x^*(T^* + \delta)) \right.$$
$$\left. - \int_{t_0}^{T^*} f(t, x^*, u^*) \, dt - \phi(T^*, x^*(T^*)) \right]$$
$$= \lim_{\delta \to 0} \frac{1}{\delta} \int_{T^*}^{T^*+\delta} f(t, x^*(t), u^*(t)) \, dt$$
$$+ \frac{\phi(T^* + \delta, x^*(T^* + \delta)) - \phi(T^*, x^*(T^*))}{\delta}$$
$$= f(T^*, x^*(T^*), u^*(T^*)) + \phi_t(T^*, x^*(T^*)) + \phi_x(T^*, x^*(T^*)) \frac{dx^*}{dt}(T^*)$$
$$= f(T^*, x^*(T^*), u^*(T^*)) + \lambda(T^*) g(T^*, x^*(T^*), u^*(T^*)) + \phi_t(T^*, x^*(T^*))$$
$$= H(T^*, x^*(T^*), u^*(T^*), \lambda(T^*)) + \phi_t(T^*, x^*(T^*)).$$

We see the need for extending u^* and x^* in the first and second lines, as the values of t considered are greater than T^*, in the case when $\delta > 0$. The transition from the second to the third line follows via the Fundamental Theorem of Calculus and the product rule. This is due to our earlier assurance that u^* is continuous at T^*, and thus x^* is differentiable at T^*.

This gives the new necessary condition we promised. Namely,

$$H(T^*, x^*(T^*), u^*(T^*), \lambda(T^*)) + \phi_t(T^*, x^*(T^*)) = 0.$$

In the case when ϕ is a function of $x(T)$ only, this says the Hamiltonian is 0 at the terminal time. This proof was done on a simplified problem for convenience. It should be clear, however, that the same necessary conditions would arise with bounds on the control and multiple states and controls as before, and that this new necessary condition would be unchanged.

What is not clear, though, is how problems with a state fixed at both endpoints are affected. Because we did not provide a development of this case, it is not obvious how a free terminal time would alter the necessary conditions. In fact, the same new necessary condition arises. Stated formally, if u^* is an optimal control on the finite, nonempty interval $[t_0, T^*]$, with control x^*, for the optimal control problem

$$\max_{u,T} \int_{t_0}^{T} f(t, x(t), u(t)) \, dt + \phi(T)$$

subject to $\quad x'(t) = g(t, x(t), u(t)), \; x(t_0) = x_0, \; x(T) = x_1,$

then for some piecewise differentiable adjoint variable λ, the following necessary conditions are simultaneously satisfied by u^*, x^*, λ:

$$0 = \frac{\partial H}{\partial u} \text{ at } u^*,$$
$$\lambda' = -\frac{\partial H}{\partial x},$$
$$0 = H(T^*, x^*(T^*), u^*(T^*), \lambda(T^*)) + \phi'(T^*).$$

Of course, the full statement should include a constant λ_0, as in Theorem 3.1, but we still assume $\lambda_0 = 1$. As before, we do not provide proofs here, as this result is a great deal more involved than the one provided. See [158].

Example 20.1 Let us return now to the problem introduced at the beginning of the chapter,

$$\min_{u,T} x(T) + \int_0^T u(t)^2 \, dt$$
$$\text{subject to} \quad x'(t) = \alpha x(t) - u(t), \; x(0) = x_0, \alpha > 0.$$

The necessary conditions

$$H = u^2 + \alpha x \lambda - u\lambda,$$
$$0 = \frac{\partial H}{\partial u} = 2u - \lambda \text{ at } u^*,$$
$$\lambda' = -\frac{\partial H}{\partial x} = -\alpha \lambda,$$
$$\lambda(T^*) = \phi_x(x^*(T^*)) = 1,$$

are exactly as in Example 3.3, so we arrive at the same solutions:

$$u^*(t) = \frac{e^{\alpha(T^*-t)}}{2},$$
$$x^*(t) = x_0 e^{\alpha t} + e^{\alpha T^*} \frac{e^{-\alpha t} - e^{\alpha t}}{4\alpha},$$
$$\lambda(t) = e^{\alpha(T^*-t)}.$$

Before, T was some fixed constant. Here, T^* is an unknown constant, which must be found using the last necessary condition:

$$0 = H(T^*, x^*(T^*), u^*(T^*), \lambda(T^*))$$
$$= \frac{1}{4} + \alpha x_0 e^{\alpha T^*} + \frac{1 - e^{2\alpha T^*}}{4} - \frac{1}{2}$$
$$= \alpha x_0 e^{\alpha T^*} - \frac{1}{4} e^{2\alpha T^*},$$

which can easily be solved to show $T^* = \frac{1}{\alpha}\ln(4\alpha x_0)$. Of course, this only makes sense if $4\alpha x_0 > 1$, which can be taken as an extra assumption on x_0. Figure 20.1 shows the optimal control.

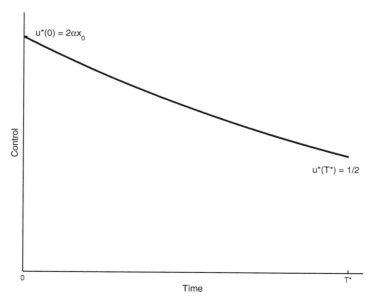

FIGURE 20.1: Optimal control from Example 20.1.

Example 20.2 (from [169])

$$\min_{u,T} \frac{1}{2}\int_0^T u(t)^2 + 1\, dt$$
subject to $\quad x'(t) = u(t),\ x(0) = 5,\ x(T) = 0,$
$$-2 \le u(t) \le 2.$$

Form the Hamiltonian

$$H = \frac{1}{2}(u^2 + 1) + \lambda u.$$

The adjoint equation is

$$\lambda' = -\frac{\partial H}{\partial x} = 0,$$

so that $\lambda \equiv c$ for some constant c. Further,

$$\frac{\partial H}{\partial u} = u + c,$$
$$0 > u^* + c \Rightarrow u^* = 2 \Rightarrow -2 > c,$$
$$0 < u^* + c \Rightarrow u^* = -2 \Rightarrow 2 < c,$$
$$0 = u^* + c \Rightarrow -2 \leq u^* \leq 2 \Rightarrow -2 \leq c \leq 2.$$

Clearly, as c is a constant, only one of these cases can be true. Thus, u^* is identically constant. Also, as the control must push the state from 5 down to 0, it is also clear that u^* must be negative. So, either $u^* \equiv -2$ or $u^* \equiv -c$.
If $u^* \equiv -2$, then the Hamiltonian is

$$H \equiv \frac{5}{2} - 2c.$$

As the terminal time is free, H must be zero at the final time. This allows us to solve for c giving $c = 5/4$. However, this contradicts what we saw above, namely, when $u^* \equiv -2$ we must have $c > 2$.

Thus, $u^* = -c$, and

$$0 = H(T^*, x^*(T^*), u^*(T^*), \lambda(T^*))$$
$$= \frac{1}{2}(u^2 + 1) + \lambda u$$
$$= \frac{1}{2}(c^2 + 1) - c^2.$$

This yields $c = \pm 1$. As the control must be negative, we have $c = 1$ and $u^* \equiv -1$. This and $x(0) = 5$ gives $x^*(t) = -t + 5$ so that $T^* = 5$.

20.2 Time Optimal Control

Of particular interest is a specific type of free terminal time problems called minimal time problems, or time optimal control. The idea is simple: move a state (or states) from a given initial location to a specified final position in minimum time. It may not be immediately clear that this confirms to the form discussed above, but note that

$$T = \int_0^T 1 \, dt.$$

Therefore, the problem

$$\min_{u,T} \int_0^T 1\,dt$$

subject to $\quad x'(t) = g(t, x(t), u(t))$, $x(0) = x_0$, $x(T) = x_1$,
$$a \leq u(t) \leq b,$$

is precisely what we want, namely, to find a control u which moves x from x_0 to x_1, subject to its dynamics, in minimal time. Of course, this is just as easily done with multiple states and controls. We make the note here that more complicated terminal state conditions or constraints can be used. Many times in applications, we are interested instead in moving the state or states from a specific initial condition to a certain region in minimal time. For example, we could only require $x(T)$ to be close to x_1, i.e., $|x(T)-x_1| \leq \delta$. Or, if we have two states, we could simply require they be equal, $x_1(T) = x_2(T)$, or be close, i.e, $|x_1(T) - x_2(T)| \leq \delta$. In general, the constraints

$$k(T, x_1(T), \ldots, x_n(T)) = 0 \quad \text{and}$$
$$k(T, x_1(T), \ldots, x_n(T)) \geq 0,$$

where k is a continuously differentiable function in all variables, can be considered. We do not treat these conditions, as such problems are generally a great deal more complicated. We refer the reader to [99, 100, 130, 141]. For examples of such problems, see [98, 120].

Example 20.3

$$\min_{u,T} \int_0^T 1\,dt$$

subject to $\quad x'(t) = x(t)u(t) - \frac{1}{2}u(t)^2$, $x(0) = x_0 \in (0,1)$, $x(T) = 1$.

We write the Hamiltonian

$$H = 1 + xu\lambda - \frac{1}{2}u^2\lambda.$$

The adjoint equation is

$$\lambda' = -\frac{\partial H}{\partial x} = -\lambda u,$$

which gives

$$\lambda(t) = C \exp\left(-\int_0^t u(s)\,ds\right),$$

for some constant C. Note, if $C = 0$, then $\lambda \equiv 0$. This gives $H \equiv 1$, which contradicts the Hamiltonian being 0 at T^*. Thus, $C \neq 0$ so that λ is never zero. Hence, the optimality condition

$$0 = \frac{\partial H}{\partial u} = \lambda(x - u)$$

gives

$$u^* = x^*.$$

Making this substitution in the state equation, we see x^* satisfies

$$x' = \frac{1}{2}x^2, \; x(0) = x_0.$$

This gives the solutions

$$x^*(t) = \frac{2x_0}{2 - x_0 t} = u^*(t).$$

The condition $x(T) = 1$ gives $T^* = 2/x_0 - 2$.

Example 20.4 Let $x(t)$ represent the location of a particle at time t. Initially, it is at rest and is positioned at $x_0 > 0$. We can steer the particle by controlling its acceleration, within its designated limits. Find the acceleration which brings x to a rest at position 0 in minimum time. Specifically,

$$\min_{u,T} \int_0^T 1 \, dt$$

subject to $x''(t) = u(t), \; x(0) = x_0 > 0, \; x(T) = 0,$
$x'(0) = 0, \; x'(T) = 0, \; -1 \leq u(t) \leq 1.$

First, we recast this as a systems problem

$$\min_{u,T} \int_0^T 1 \, dt$$

subject to $x_1'(t) = x_2(t), \; x_1(0) = x_0 > 0, \; x_1(T) = 0,$
$x_2'(t) = u(t), \; x_2(0) = 0, \; x_2(T) = 0,$
$-1 \leq u(t) \leq 1.$

The Hamiltonian is

$$H = 1 + \lambda_1 x_2 + \lambda_2 u.$$

From the adjoint equations

$$\lambda_1' = -\frac{\partial H}{\partial x_1} = 0,$$
$$\lambda_2' = -\frac{\partial H}{\partial x_2} = -\lambda_1,$$

it is clear λ_1 is identically some constant, and λ_2 is a linear function in t. If λ_2 were identically 0, then $\lambda_1 \equiv \lambda_2 \equiv 0$, from which we see $H \equiv 1$. This contradicts $H(T^*) = 0$, so λ_2 is not identically 0. Further,

$$\frac{\partial H}{\partial u} = \lambda_2.$$

As λ_2 is linear and not identically 0, it can be 0 only at a point and only once, so that u^* is bang-bang with at most one switch. Now, $x_2' = u$ and x_2 is to begin and end at 0. Therefore, it is clear that u cannot be identically -1 or 1, but must utilize the one allowed switch. It should also be clear that this switch occurs at the half-way point of the interval, $T^*/2$. The only thing to determine is which bound u^* begins with.

Suppose the optimal control is

$$u^*(t) = \begin{cases} 1 & \text{when} \quad 0 \leq t < T^*/2, \\ -1 & \text{when} \quad T^*/2 < t \leq T^*. \end{cases}$$

Using the state equation and $x_2(0) = 0 = x_2(T^*)$, we can see

$$x_2^*(t) = \begin{cases} t & \text{when} \quad 0 \leq t \leq T^*/2, \\ T^* - t & \text{when} \quad T^*/2 \leq t \leq T^*. \end{cases}$$

Using $x_1' = x_2$, $x_1(0) = x_0$, and $x_1(T^*) = 0$, it follows

$$x_1^*(t) = \begin{cases} \frac{1}{2}t^2 + x_0 & \text{when} \quad 0 \leq t \leq T^*/2, \\ -\frac{1}{2}t^2 + T^*t - \frac{1}{2}(T^*)^2 & \text{when} \quad T^*/2 \leq t \leq T^*. \end{cases}$$

Now, x_1^* is continuous, so the two expressions must agree at $T^*/2$. This implies

$$\frac{1}{2}(T^*/2)^2 + x_0 = -\frac{1}{2}(T^*/2)^2 + T^*(T^*/2) - \frac{1}{2}(T^*)^2 \Rightarrow$$
$$x_0 = -(T^*/2)^2 < 0.$$

This contradicts the original assumption of x_0. Therefore, the optimal control, and resulting optimal states, must be

$$u^*(t) = \begin{cases} -1 & \text{when} \quad 0 \le t < T^*/2, \\ 1 & \text{when} \quad T^*/2 < t \le T^*, \end{cases}$$

$$x_1^*(t) = \begin{cases} -\frac{1}{2}t^2 + x_0 & \text{when} \quad 0 \le t \le T^*/2, \\ \frac{1}{2}t^2 - T^*t + \frac{1}{2}(T^*)^2 & \text{when} \quad T^*/2 \le t \le T^*, \end{cases}$$

$$x_2^*(t) = \begin{cases} -t & \text{when} \quad 0 \le t \le T^*/2, \\ t - T^* & \text{when} \quad T^*/2 \le t \le T^*. \end{cases}$$

Using the fact that x_1^* must be continuous, we find $x_0 = (T^*/2)^2$, so that $T^* = 2\sqrt{x_0}$. This can be substituted into the expressions above to finish the problem. The optimal states are shown in Figure 20.2.

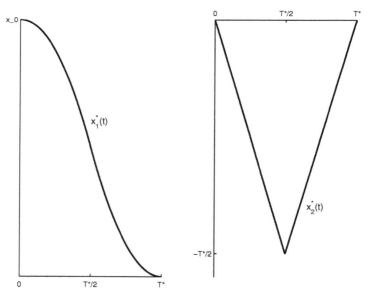

FIGURE 20.2: The optimal states for Example 20.4.

20.3 Exercises

Exercise 20.1 (from [130]) Solve

$$\min_{u,T} \int_0^T 1 + \frac{1}{2}u(t)^2 \, dt$$

subject to $x_1'(t) = x_2(t)$, $x_1(0) = 1$, $x_1(T) = 0$,
$x_2'(t) = u(t)$, $x_2(0) = 0$,
$-1 \leq u(t) \leq 1$.

Exercise 20.2 (from [126]) Let $x_0 > x_1 > 0$. Solve

$$\min_{u,T} \frac{1}{2} \int_0^T u(t)^2 \, dt$$

subject to $x'(t) = x(t) + u(t)$, $x(0) = x_0$, $x(T) = x_1$.

Exercise 20.3 (from [126]) Let $0 < x_0 < 1$. Solve

$$\min_{u,T} \int_0^T 1 \, dt$$

subject to $x'(t) = x(t) - u(t)$, $x(0) = x_0$, $x(T) = 0$,
$-1 \leq u(t) \leq 1$.

Exercise 20.4 (from [141]) Solve

$$\min_{u,T} \int_0^T 1 \, dt$$

subject to $x_1'(t) = x_2(t)$, $x_1(0) = x_0$, $x_1(T) = 0$,
$x_2'(t) = -x_1(t) + u(t)$, $x_2(T) = 0$,
$-1 \leq u(t) \leq 1$.

Chapter 21

Adapted Forward-Backward Sweep

The forward-backward sweep method, as used so far, is somewhat limited in the type of optimal control problems it can solve. In this chapter, we develop a modified version of the forward-backward sweep, which will allow us to examine more complicated problems.

21.1 Secant Method

Before moving forward with this new method to solve optimal control problems, it is advantageous to first introduce a basic algorithm for finding zeros of functions, the secant method. For those with some familiarity of numerical methods, the secant method is simply Newton's method, where the derivative is replaced by a difference quotient. We give a brief development of the method here. For more information, see [32].

Let V be a real-valued function. Suppose we are interested in the zeros of V, i.e., the values x for which $V(x) = 0$. The idea is to form a sequence of values or *nodes* $x_1, x_2, \ldots$ which converge to such a zero. We take x_1, x_2 to be given and not equal. We form the sequence by induction.

Suppose we have $x_1, \ldots, x_n$. Further, suppose $V'(x_n)$ is well-defined. Consider the linear approximation of V at x_n, given by $L(x) = V'(x_n)(x - x_n) + V(x_n)$. We set x_{n+1} to be the zero of $L(x)$. Namely,

$$x_{n+1} = x_n - \frac{V(x_n)}{V'(x_n)}.$$

This is precisely the idea of Newton's method. If V' is defined everywhere, then this sequence will, in fact, converge to a zero of V.

However, we are interested in the case when V' may not be known. For small h, we have

$$V'(x) \approx \frac{V(x+h) - V(x)}{h}.$$

If we set $x = x_n$ and $h = x_{n-1} - x_n$, then

$$V'(x_n) \approx \frac{V(x_{n-1}) - V(x_n)}{x_{n-1} - x_n}.$$

Substituting this into the formula above, we arrive at

$$x_{n+1} = x_n - \left(\frac{x_n - x_{n-1}}{V(x_n) - V(x_{n-1})}\right) V(x_n).$$

By induction, we can find the nodes x_n for all n. The idea is to continue construction of the sequence until V is "close" to 0, i.e., $|V(x_n)| < \epsilon$, for some error tolerance ϵ. A pseudocode for a secant method in MATLAB is provided below.

─────────────── secantcode.m ───────────────
```
function y = secantcode(a,b,epsilon)

flag = -1;

Va = V(a);
Vb = V(b);

while(flag < 0)
    if(abs(Va) > abs(Vb))
        k = a;
        a = b;
        b = k;
        k = Va;
        Va = Vb;
        Vb = k;
    end

    d = Va*(b - a)/(Vb - Va);
    b = a;
    Vb = Va;
    a = a - d;
    Va = V(a);

    if(abs(Va) < epsilon)
        flag = 1;
    end
end

y = a;
```

The letters a and b are used throughout the code for the two current nodes in the sequence, while Va and Vb are used to represent the values of V at a,

b. As we are not interested in the actual sequence, only the last element, we abandon the notation x_n all together, and use a, b as inputs. Lines 5 and 6 calculate the function value at our initial a, b. Line 8 begins the *while* loop, which will calculate the successive nodes. The variable *flag*, as before, acts as the on-off switch for the loop. Lines 9 through 16 switch the values of a, b, and Va, Vb if $|Va| > |Vb|$. This is done to keep the "better" value (the one closer to zero) and replace the other. The variable k is just a temporary place holder here. Line 18 calculates the difference quotient times the function value. Lines 19 and 20 store the new values as the old values, and lines 21 and 22 calculate the new values. Line 24 checks how close our function value is to 0, while line 29 outputs the last known element of the sequence, once the error is small enough. Note, in lines 5, 6, and 22, the use of $V(a)$ and $V(b)$ only makes sense if $V(\cdot)$ is defined in MATLAB. Specifically, V must be an inline function or a function defined by another code, as will be done later.

21.2 One State with Fixed Endpoints

Consider the optimal control problem

$$\max_{u_1,\ldots,u_m} \int_{t_0}^{t_1} f(t, x_1(t), \ldots, x_n(t), u_1(t), \ldots, u_m(t))\, dt$$

subject to $\quad x_1'(t) = g_1(t, x_1(t), \ldots, x_n(t), u_1(t), \ldots, u_m(t)),\ x_1(t_0) = x_{10},$
$\qquad\qquad\ \ x_2'(t) = g_2(t, x_1(t), \ldots, x_n(t), u_1(t), \ldots, u_m(t)),\ x_2(t_0) = x_{20},$
$\qquad\qquad\ \ \vdots$
$\qquad\qquad\ \ x_n'(t) = g_n(t, x_1(t), \ldots, x_n(t), u_1(t), \ldots, u_m(t)),\ x_n(t_0) = x_{n0},$
$\qquad\qquad\ \ x_n(t_1) = x_{n1},$
$\qquad\qquad\ \ a_j \leq u_j(t) \leq b_j \ \text{ for } j = 1, 2, \ldots, m.$

Note, all states are free at the terminal time, except x_n, which is fixed at both endpoints. As we stated earlier, our current numerical methods cannot deal with this problem. Let us now develop a way to solve it. We will have n adjoints, $\lambda_1, \ldots, \lambda_n$. Now, $\lambda_j(t_1) = 0$ for $j = 1, 2, \ldots, n-1$, but $\lambda_n(t_1)$ is unknown.

Suppose we make a guess $\lambda_n(t_1) = \theta$. Then, construct a forward-backward sweep code to solve this problem, using the given initial conditions and $\vec{\lambda}(t_1) = (0, 0, \ldots, 0, \theta)$. Assuming convergence occurs, we will be given a value for the nth state at the terminal time. Denote this estimate by $\widetilde{x}_{n1}$. It is unlikely our guess yielded the correct value; in particular, $\widetilde{x}_{n1} \neq x_{n1}$. However, we can consider the map $\theta \mapsto \widetilde{x}_{n1}$ as a function. Then, our problem boils down to finding the θ value for which $\widetilde{x}_{n1}$ is actually x_{n1}. Or, equivalently, if we define

$$V(\theta) = \tilde{x}_{n1} - x_{n1},$$

we are interested in the zeros of V.

To carry this out requires the construction of two separate codes. The first code will be a forward-backward sweep routine, which takes as input the guess θ. The second code will be the secant code provided above, with V replaced by the first code. This is essentially a type of shooting method. For this text, we refer to it as the Adapted Forward-Backward Sweep method. We illustrate the idea with an example.

Example 21.1

$$\min_u \frac{1}{2} \int_0^1 x_1(t)^2 + u(t)^2 \, dt$$

subject to $x_1'(t) = x_2(t)$, $x_1(0) = 1$,
$x_2'(t) = -ru(t)$, $x_2(0) = 0$, $x_2(1) = s$,
$r > 0$, $s < 0$.

Note, our function of interest here is $V(\theta) = \tilde{x}_{21} - s$. We write two codes for this problem, *example1.m* and *example2.m*. The code *example1.m* solves the above problem, ignoring the condition $x_2(1) = s$ and guessing $\lambda_2(1) = \theta$. The code *example2.m* is the secant method to find the appropriate 0.

――――――――――――――――――――― example1.m ―――――――――――――――――――――
```
1   function y = example1(theta,r)
2
3   test = -1;
4
5   delta = 0.001;
6   N = 1000;
7   t=linspace(0,1,N+1);
8   h=1/N;
9   h2 = h/2;
10
11  x1=zeros(1,N+1);
12  x1(1)=1;
13  x2=zeros(1,N+1);
14  x2(1)=0;
15
16  lambda1=zeros(1,N+1);
17  lambda2=zeros(1,N+1);
18  lambda2(N+1) = theta;
19
20  u=zeros(1,N+1);
21
22  while(test < 0)
```
――――――――――――――――――――― example1.m ―――――――――――――――――――――
```
74  y(1,:)=t;
75  y(2,:)=x1;
76  y(3,:)=x2;
77  y(4,:)=u;
```

The missing portion of the code is simply the forward and backward sweeps, the update of u, and convergence tests, all as usual. For the most part, this is the standard forward-backward sweep we have been using. Line 16 sets $\lambda_2(1)$ to θ. Also, as this first portion ignores the condition $x_2(1) = s$, the variable s is not an input of *example1.m*.

```
                        ──────── example2.m ────────
 1    function y = example2(a,b,r,s)
 2
 3    flag = -1;
 4
 5    z = example1(a,r);
 6    Va = z(3,1001) - s;
 7    z = example1(b,r);
 8    Vb = z(3,1001) - s;
 9
10    while(flag < 0)
11        if(abs(Va) > abs(Vb))
12            k = a;
13            a = b;
14            b = k;
15            k = Va;
16            Va = Vb;
17            Vb = k;
18        end
19
20        d = Va*(b - a)/(Vb - Va);
21        b = a;
22        Vb = Va;
23        a = a - d;
24        z = example1(a,r);
25        Va = z(3,1001) - s;
26
27        if(abs(Va) < 1E-10)
28            flag = 1;
29        end
30    end
31
32    y = z;
```

First, note that the error tolerance ϵ has been set to 10^{-10}. Also, the variables r and s are inputs. In line 5, the forward-backward sweep of *example1.m* is run, with the initial guess $\lambda_2(1) = a$ and the stored value for r. Once the forward-backward sweep converges, the values are stored as z, a 4×1001 array containing the values for t, x_1, x_2, and u. In line 6, $V(a)$ is found, as $z(3, 1001)$ is the stored value for $x_2(1)$. The same thing occurs in lines 7, 8 and 24, 25. Finally, instead of outputting only the 0 of the function, as was done in *secantcode.m*, here we output the stored values for z. This gives the values for the variables t, x_1, x_2, u during the last iteration, when V was nearly 0. Therefore, we take these to be the solutions of the optimal control problem.

Adapted Forward-Backward Sweep

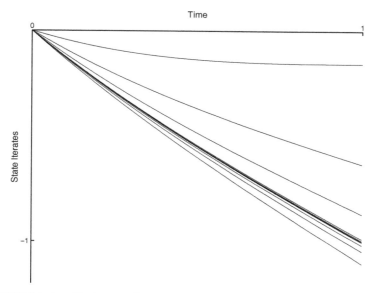

FIGURE 21.1: Estimates for x_2^* in Example 21.1. Note, the iterates cluster to the correct function, which has value -1 at $t = 1$.

Figure 21.1 shows the iterations of this method for Example 21.1, for $r = 1$ and $s = -1$. For each guess θ, a full forward-backward sweep runs and must converge. After this, we have an estimate for x_2, all of which are shown in the figure together. Once $x_2(1)$ is close enough to $s = -1$, the program stops.

This method is of particular interest when dealing with optimal control problems involving isoperimetric constraints. Recall from section 12.4, one method of solving such problems is to introduce an artificial state variable, which is fixed at both endpoints. This numerical method allows us to deal with such problems. Lab 13 focuses on a biological model with an isoperimetric constraint.

The adapted forward-backward sweep method is based on two critical assumptions: the forward-backward sweep will converge for most choices of θ, and the function V is well-defined. For "well-behaved" problems, both are usually true. There, the choices of a, b (the beginning nodes) are usually irrelevant. Most $a \neq b$ will arrive at the same set of solutions. Of course, guesses which are closer to the actual values generally lead to quicker convergence. On the other hand, for certain ill-behaved problems, these assumptions may not hold. Many choices of θ can lead to difficulty with convergence, while V may be a well-defined function only on some interval, not the whole real line. In this case, the choice of a, b is critical, and experimentation must be performed. This is why a, b are left as inputs in our code, and not set to universal values.

We would also like to point out that any root-finding algorithm will work in place of the secant method. Indeed, for problems where the secant-based method has trouble converging, a bisection-based method, although much slower, is a good alternative. For more information on basic root-finding methods, see [32].

21.3 Nonlinear Payoff Terms

A slight alteration of the adapted forward-backward sweep can be used to numerically solve optimal control problems with nonlinear payoff terms. Consider the problem

$$\max_{u_1,\ldots,u_m} \phi(x_n(t_1)) + \int_{t_0}^{t_1} f(t, x_1(t), \ldots, x_n(t), u_1(t), \ldots, u_m(t))\, dt$$

subject to
$$x_1'(t) = g_1(t, x_1(t), \ldots, x_n(t), u_1(t), \ldots, u_m(t)), \quad x_1(t_0) = x_{10},$$
$$x_2'(t) = g_2(t, x_1(t), \ldots, x_n(t), u_1(t), \ldots, u_m(t)), \quad x_2(t_0) = x_{20},$$
$$\vdots$$
$$x_n'(t) = g_n(t, x_1(t), \ldots, x_n(t), u_1(t), \ldots, u_m(t)), \quad x_n(t_0) = x_{n0},$$
$$a_j \leq u_j(t) \leq b_j \quad \text{for } j = 1, 2, \ldots, m.$$

Now, if ϕ is linear, i.e., $\phi' \equiv c$ for some constant c, then the standard forward-backward sweep can be used. Simply set $\lambda_j(t_1) = 0$ for $1 \leq j \leq n-1$ and $\lambda_n(t_1) = c$. However, for any other ϕ, the value $\lambda_n(t_1)$ will not be constant, but will depend on $x_n^*(t_1)$. In this case, the standard forward-backward sweep cannot be used, but the adapted method can be applied.

As before, we make a guess for the value of adjoint at the terminal time, say $\lambda_n(t_1) = \theta$. Then, we use the standard forward-backward sweep to solve the problem with given initial conditions and $\lambda_n(t_1) = \theta$. This will give a value for the nth state at t_1, say $\tilde{x}_{n1}$. We want $\phi'(\tilde{x}_{n1}) = \theta$. So, consider the map $\theta \mapsto \tilde{x}_{n1}$, and define a function $V(\theta) = \phi'(\tilde{x}_{n1}) - \theta$. Thus, our problem has been reduced to finding the zeros of V. The secant code given earlier can now be employed. Let us follow with an example.

Example 21.2

$$\min_u \frac{1}{4}x_2(1)^4 + \frac{1}{2}\int_0^1 x_1(t)^2 + u(t)^2\, dt$$

$$\text{subject to} \quad x_1'(t) = x_2(t),\ x_1(0) = 1,$$
$$x_2'(t) = -ru(t),\ x_2(0) = 0,$$
$$r > 0.$$

First, note this example is nearly identical to Example 21.1, the only difference being $x_2(1)$ is no longer fixed, but is in the objective functional. Also, $\phi(s) = \frac{1}{4}s^4$, so that $\phi'(s) = s^3$. Thus, the function we want is $V(\theta) = (\widetilde{x}_{21})^3 - \theta$.

The first code which must be constructed is the forward-backward sweep which solves the problem given $\lambda_2(1) = \theta$; but, this is precisely *example1.m*. Nothing in the code needs to be altered. The second code is the secant method routine which will find the zeros of our V. It can easily be recovered from *example2.m*. Lines 6, 8, and 25 should be changed to, respectively,

$$Va = z(3,1001)^3 - a,$$
$$Vb = z(3,1001)^3 - b,$$
$$Va = z(3,1001)^3 - a.$$

This is done to reflect the new definition of V.

21.4 Free Terminal Time

Finally, we can apply this numerical method to non-autonomous problems which have free terminal time. Consider the problem

$$\max_{u_1,\ldots,u_m,T} \int_{t_0}^{T} f(t, x_1(t), \ldots, x_n(t), u_1(t), \ldots, u_m(t))\, dt$$

$$\text{subject to} \quad x_1'(t) = g_1(t, x_1(t), \ldots, x_n(t), u_1(t), \ldots, u_m(t)),\ x_1(t_0) = x_{10},$$
$$x_2'(t) = g_2(t, x_1(t), \ldots, x_n(t), u_1(t), \ldots, u_m(t)),\ x_2(t_0) = x_{20},$$
$$\vdots$$
$$x_n'(t) = g_n(t, x_1(t), \ldots, x_n(t), u_1(t), \ldots, u_m(t)),\ x_n(t_0) = x_{n0},$$
$$a_j \leq u_j(t) \leq b_j \text{ for } j = 1, 2, \ldots, m.$$

This time, make a guess for the optimal terminal time T^*, say $T^* = \theta > t_0$. Solve the above problem as a fixed time problem using the standard forward-backward sweep method. This will give estimates of the controls, states, and adjoints at the final time. These can be used to calculate the estimate of the Hamiltonian at the final time, which we denote $\widetilde{H}(\theta)$. If we consider the map $\theta \mapsto \widetilde{H}(\theta)$, we can define our function as $V(\theta) = \widetilde{H}(\theta)$. As we require $H(T^*) = 0$, we can simply look for the zeros of V as before.

Note, the requirement that the original problem be non-autonomous is crucial. Otherwise, the Hamiltonian would be 0 for all t.

When employing this method, one should always take the initial nodes a, b to be strictly greater than the initial time. Of the three types of problems we have discussed, free terminal time problems tend to have the longest convergence time with this method. In general, it is also free terminal time problems that have the most trouble converging. This is caused by the Hamiltonian function, which is usually more complicated than the V functions associated with fixed states and nonlinear payoff problems. Nevertheless, it is still effective enough to mention here and always worth trying for relevant problems.

21.5 Multiple Shots

So far, we have developed the Adapted forward-backward Sweep to deal with problems with one state fixed, a nonlinear payoff term, or free terminal time, but not problems with several such elements. Expansion to these kinds of problems is actually relatively simple. It is done by employing our shooting method multiple times, or doing multiple shots.

To use the basic forward-backward sweep, we require the information of the necessary conditions to be in certain locations, namely, fixed initial conditions for the states and fixed terminal conditions for the adjoints. When dealing with other problems, we may not have these conditions, but we gain other necessary conditions. For example, in problems with a nonlinear payoff term, we lose the fixed terminal condition of an adjoint, but gain a condition linking the terminal values of the adjoint and its state. The idea of the method developed in this chapter is to guess a value for the missing necessary condition and use a root-finder to determine the appropriate guess to recover this new necessary condition.

Stated another way, we ignore troublesome necessary conditions and guess other values so the problem takes on a form we are comfortable with. We solve this problem (numerous times, if necessary), until the neglected necessary condition is satisfied. Now, suppose a problem had two such necessary conditions. Then, we could put aside the first and make an appropriate guess. The resulting problem would have only one troublesome necessary, and we know how to solve this type of problem; use the Adapted Forward-Backward Sweep as just developed. This allows us to solve problems with two such elements. However, now three-element problems can be reduced to two-element problems to be solved, and so on. By induction, any multiple element problem can be solved.

For the sake of exposition, we study a problem with two states, both fixed at the endpoints. Consider

$$\max_u \int_{t_0}^{t_1} f(t, x_1(t), x_2(t), u(t))\, dt$$

subject to $x_1'(t) = g_1(t, x_1(t), x_2(t), u(t))$, $x_1(t_0) = x_{10}$, $x_1(t_1) = x_{11}$,
$x_2'(t) = g_2(t, x_1(t), x_2(t), u(t))$, $x_2(t_0) = x_{20}$, $x_1(t_1) = x_{21}$.

After writing out the necessary conditions, suppose we arrive at the system

$$\begin{aligned}
x_1'(t) &= g_1(t, x_1(t), x_2(t), u(t)), \quad x_1(t_0) = x_{10}, \ x_1(t_1) = x_{11}, \\
x_2'(t) &= g_2(t, x_1(t), x_2(t), u(t)), \quad x_2(t_0) = x_{20}, \ x_2(t_1) = x_{21}, \\
\lambda_1'(t) &= h_1(t, x_1(t), x_2(t), \lambda_1(t), \lambda_2(t), u(t)), \\
\lambda_2'(t) &= h_2(t, x_1(t), x_2(t), \lambda_1(t), \lambda_2(t), u(t)), \\
u(t) &= k(t, x_1(t), x_2(t), \lambda_1(t), \lambda_2(t)).
\end{aligned} \quad (21.1)$$

Make the guess $\lambda_1(t_1) = \theta_1$ and ignore $x_1(t_1) = x_{11}$. Then, (21.1) becomes

$$\begin{aligned}
x_1'(t) &= g_1(t, x_1(t), x_2(t), u(t)), \quad x_1(t_0) = x_{10}, \\
x_2'(t) &= g_2(t, x_1(t), x_2(t), u(t)), \quad x_2(t_0) = x_{20}, \ x_2(t_1) = x_{21}, \\
\lambda_1'(t) &= h_1(t, x_1(t), x_2(t), \lambda_1(t), \lambda_2(t), u(t)), \quad \lambda_1(t_1) = \theta_1, \\
\lambda_2'(t) &= h_2(t, x_1(t), x_2(t), \lambda_1(t), \lambda_2(t), u(t)), \\
u(t) &= k(t, x_1(t), x_2(t), \lambda_1(t), \lambda_2(t)).
\end{aligned} \quad (21.2)$$

We have discussed solving systems like (21.2). Make a guess $\lambda_2(t_1) = \theta_2$ and ignore $x_2(t_1) = x_{21}$. Then, we have

$$\begin{aligned}
x_1'(t) &= g_1(t, x_1(t), x_2(t), u(t)), \quad x_1(t_0) = x_{10}, \\
x_2'(t) &= g_2(t, x_1(t), x_2(t), u(t)), \quad x_2(t_0) = x_{20}, \\
\lambda_1'(t) &= h_1(t, x_1(t), x_2(t), \lambda_1(t), \lambda_2(t), u(t)), \quad \lambda_1(t_1) = \theta_1, \\
\lambda_2'(t) &= h_2(t, x_1(t), x_2(t), \lambda_1(t), \lambda_2(t), u(t)), \quad \lambda_2(t_1) = \theta_2, \\
u(t) &= k(t, x_1(t), x_2(t), \lambda_1(t), \lambda_2(t)).
\end{aligned} \quad (21.3)$$

This system can be solved by the standard forward-backward sweep method. For θ_1 fixed, solve (21.3) for various θ_2 using a root-finder to determine the θ_2 which achieves $x_2(t_1) = x_{21}$. Then, we have solved (21.2) for some fixed θ_1. We do this for several different θ_1. For each θ_1, we get an estimate for $x_1(t_1)$. Using a root-finder, we determine the θ_1 that yields $x_1(t_1) = x_{11}$. Note, to solve (21.1) in this manner will require solving (21.2) many times, and each time we solve (21.2) requires solving (21.3) many times.

Example 21.3

$$\min_u \frac{1}{2} \int_0^1 x_1(t)^2 + u(t)^2 \, dt$$

subject to $x_1'(t) = x_2(t)$, $x_1(0) = 1$, $x_1(1) = 0$,
$x_2'(t) = -ru(t)$, $x_2(0) = 0$, $x_2(1) = s$,
$r > 0$, $s < 0$.

Again, this is almost Example 21.1, except now both x_1, x_2 are fixed at the terminal time. Neither λ_1, λ_2 are known at 1, so we make guesses for both. The first code is the forward-backward sweep, with both θ_1, θ_2 as inputs. This can be written by taking *example1.m* and making the following changes:

———————— example1.m ————————
```
1   function y = example1(theta1,theta2,r)
```

———————— example1.m ————————
```
16  lambda1=zeros(1,N+1);
17  lambda1(N+1) = theta1;
18  lambda2=zeros(1,N+1);
19  lambda2(N+1) = theta2;
```

The second code should find the zeros of the function $V(\theta_2) = \tilde{x}_{21} - s$, with $\lambda_1(1) = \theta_1$ fixed. This is nearly *example2.m*, except we must add θ_1 as an input and reflect the changes in *example1.m*:

———————— example2.m ————————
```
1   function y = example2(theta1,a,b,r,s)
2
3   flag = -1;
4
5   z = example1(theta1,a,r);
6   Va = z(3,N+1) - s;
7   z = example1(theta1,b,r);
8   Vb = z(3,N+1) - s;
```

———————— example2.m ————————
```
24      z = example1(theta1,a,r);
25      Va = z(3,N+1) - s;
```

The third code, *example3.m*, will use the new *example2.m* above to find the zeros of the function $V(\theta_1) = \tilde{x}_{11}$. This is done by amending *example2.m* appropriately.

Adapted Forward-Backward Sweep

```
──────────────── example3.m ────────────────
1  function y = example3(a,b,c,d,r,s)
2
3  flag = -1;
4
5  z = example2(a,c,d,r,s);
6  Va = z(2,N+1);
7  z = example2(b,c,d,r,s);
8  Vb = z(2,N+1);
```

```
──────────────── example3.m ────────────────
24     z = example2(a,c,d,r,s);
25     Va = z(2,N+1);
```

Here, a, b are the beginning nodes for the x_1-secant routine, and c, d are the beginning nodes for x_2-secant routine.

This method is of great use for problems containing multiple fixed states and/or nonlinear payoff terms. However, due to the discussed difficulty with free terminal time problems, it has proven generally ineffective for multiple element problems where the terminal time is also free. For example, minimum time problems, in addition to having free terminal time, usually contain a fixed state or states. While we would like to employ this method on such problems, it often fails. Other methods, usually quite complicated, are needed; see [12].

21.6 Exercises

Exercise 21.1 Reconsider the bioreactor problem of Lab 12:

$$\min_u \ \ln(z(T)) + \int_0^T Au(t)\,dt$$

$$\text{subject to} \quad x'(t) = Gu(t)x(t) - Dx(t)^2, \ x(0) = x_0,$$
$$z'(t) = -Kx(t)z(t), \ z(0) = z_0,$$
$$0 \le u(t) \le M.$$

Instead of making the simplification done in Lab 12, solve the problem exactly as written. In particular, use the Adapted Forward-Backward Sweep to solve this problem numerically. Verify that you get the same solutions as given by *lab12*.

Exercise 21.2 Write a code to solve the following problem:

$$\min_u \frac{1}{2} \int_0^1 u(t)^2\, dt$$

subject to $x_1'(t) = x_2(t)$, $x_1(0) = 0$, $x_1(1) = 1$,
$x_2'(t) = u(t)$, $x_2(0) = 0$, $x_2(1) = 0$.

Solve the problem analytically and verify the solutions are the same.

Chapter 22

Lab 13: Predator-Prey Model

In this lab, we study a simple predator-prey model which contains an isoperimetric constraint. It is partially based on the work by Goh, Leitmann, and Vincent [70]. We start with the standard Lotka-Volterra model

$$N_1'(t) = (\alpha_1 - \beta_1 N_2(t))N_1(t), \; N_1(0) = N_{10},$$
$$N_2'(t) = (\beta_2 N_1(t) - \alpha_2)N_2(t), \; N_2(0) = N_{20},$$

where $N_1(t)$ is the prey population at time t, and $N_2(t)$ is the predator population. Here, we have scaled time to some arbitrary unit. Also, α_1, α_2, β_1, and β_2 are positive constants, subject to this time scaling.

We wish to consider a situation where the prey act as a pest, such as an insect population. The goal should be to reduce the pest population with the use of a chemical or biological agent, or pesticide. An ideal pesticide is one that affects only the pests (not the predators), leaves no residue, and kills in a density dependent manner. In practice, none of these is usually true. For simplicity, we study a pesticide which adheres to the last two assumptions. More information on models with less ideal pesticides can be found in [70, 81, 180].

Suppose the application of a pesticide kills both the pest/prey and predator in a density dependent manner, with density parameters $d_1 > 0$ and $d_2 > 0$ respectively. Let $u(t)$ be the rate of application at time t. Then, our model becomes

$$N_1'(t) = (\alpha_1 - \beta_1 N_2(t))N_1(t) - d_1 N_1(t)u(t), \; N_1(0) = N_{10} > 0,$$
$$N_2'(t) = (\beta_2 N_1(t) - \alpha_2)N_2(t) - d_2 N_2(t)u(t), \; N_2(0) = N_{20} > 0.$$

Our goal should be to minimize the pest population at some specified time T. However, the levels of pesticide should also be taken into account. Suppose environmental and/or economic restrictions allow only a certain fixed level of application at any given time (say M) and only a certain fixed total application over the time period (say B). Finally, for simplicity, we take $\alpha_1 = \alpha_2 = \beta_1 = \beta_2 = 1$. Then, our optimal control problem can be cast as

$$\min_{u} \; N_1(T) + \frac{A}{2} \int_0^T u(t)^2 \, dt$$

subject to $N_1'(t) = (1 - N_2(t))N_1(t) - d_1 N_1(t)u(t)$, $N_1(0) = N_{10}$,
$N_2'(t) = (N_1(t) - 1)N_2(t) - d_2 N_2(t)u(t)$, $N_2(0) = N_{20}$,

$$0 \leq u(t) \leq M, \qquad \int_0^T u(t)\,dt = B.$$

Note $B = \int_0^T u(t)\,dt \leq \int_0^T M\,dt = MT$. So, the constants B, M, and T cannot be chosen completely independent of each other. This compatibility condition is enforced in the code provided.

Observe, the integral of u over the time period is fixed, while the integral of u^2 is included in the objective functional. This may be somewhat puzzling, but is done for two reasons. First, while the total amount of pesticide is fixed, the integral of u^2 can vary. Indeed, this objective functional penalizes overly large values of u and encourages smaller values. Environmentally, we can see why such a consideration is made. The second reason is a purely practical one. The introduction of the u^2 term prevents our problem from being linear in the control, and our usual techniques can be applied. However, if large amounts of pesticide are of no concern, the weight parameter A can be taken exceedingly small to limit the effect this term has.

To solve this problem numerically, we introduce the state variable $z(t)$ with $z'(t) = u(t)$, $z(0) = 0$, and $z(T) = B$. The adapted forward-backward sweep can then be used to solve this three-state problem, which is precisely what is done. The two pieces involved, the standard forward-backward sweep routine which takes $\lambda_3(T)$ as an input and the accompanying secant routine, have been placed in one m-file *code13.m* for convenience. There are also a few cosmetic alterations to the code developed in the previous chapter. You will now be asked for the initial nodes a and b of $\lambda_3(T)$. In the simulations spelled out here, nodes are provided which will lead to speedy convergence. The code also updates you on how many iterations of the forward-backward sweep have been implemented. Finally, the last guess of $\lambda_3(T)$ will be given, which may be helpful when you run your own simulations. The program *lab13.m* is the usual interface.

Finally, the reader should keep in mind that this method is highly sensitive. As discussed in the previous chapter, the choice of nodes a and b can be critical, and the best choices will of course vary with the constants in the problem. Certain constant choices will have trouble converging, and it may take several tries to find the "correct" a and b. Still other choices may lead to a situation where convergence is impossible. When you run simulations with your own data, watch the iterations. Each iteration should occur relatively quickly: about 5 seconds, no more than 10. And overall convergence should occur in no more than 15 iterations. If it goes beyond this, you might want to consider supplying different nodes.

We are now ready to begin this lab. Open MATLAB and enter *lab13*. Begin with the values

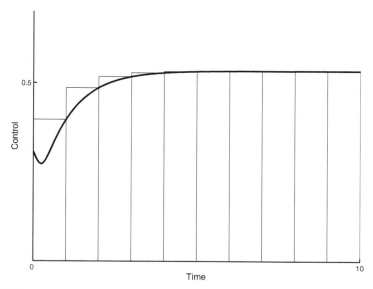

FIGURE 22.1: Optimal control from the (22.1) simulation. Here, the right-hand sum approximation is 5.1349.

$$\boxed{\begin{array}{llllll} d_1 = 0.1 & d_2 = 0.1 & N_{10} = 10 & N_{20} = 1 & M = 1 \\ A = 1 & B = 5 & T = 10 & a = -0.52 & b = -0.5 \end{array}} \quad . \tag{22.1}$$

Do not vary any parameters yet. The optimal control is shown in Figure 22.1. In this simulation, both the prey and predator populations are nearly eliminated. There is an initial rise in predator population, as the prey population is large enough to support an increase. However, as the pesticide regimen strengthens and the prey population declines, so does the predator. Here, $d_1 = d_2$, which is not realistic. We would not use a pesticide that eroded the predator and prey populations at the same rate. Also, the total pesticide level B is likely unrealistically high.

Try the values

$$\boxed{\begin{array}{llllll} d_1 = 0.1 & d_2 = 0.01 & N_{10} = 5 & N_{20} = 2 & M = 1 \\ A = 1 & B = 1 & T = 5 & a = -0.2 & b = -0.18 \end{array}} \quad . \tag{22.2}$$

Even with these changes in d_2, N_{10}, N_{20}, B, T, a, and b, we still have a similar result. The pesticide regimen, while generally weaker than before, still starts at its minimum value and increases, becoming nearly constant for much of the time period. The predator population initially increases, and both populations are nearly wiped out. Perhaps we are still being too flippant about the negative effects of the pesticide, when we use such a low value of A. Enter the (22.2) values again, this time varying with $A = 10$. The total value of pesticide is fixed, so the change in A only affects the overall distribution of the pesticide. In particular, the initial values are raised and the later values lowered so that the second control is nearly constant the entire time interval. However, something unusual occurs to the two populations: they do not change at all. Despite the fact that a new pesticide regimen is in place, the prey and predator populations experience no change.

You will recall in the previous labs we have touched on this duality. Namely, when certain constants are changed, there are two ways the new optimal control can be affected: the control is altered in order to achieve the same state results, more or less, or the control is unchanged, while the state systems are shifted. More often than not, the new optimal control was a compromise between these two scenarios. On occasion, though, we have pointed out systems which leaned more heavily in one direction. This is a relatively extreme example of the first option. In fact, this behavior is not limited to changes in A. Indeed, it is seen in virtually all the constants.

Enter (22.2) and vary with $d_1 = 0.3$. Again, the control is changed (as expected) but the populations remain the same. Vary each of d_2, M, and B in turn (making sure the compatibility condition $B \leq MT$ is satisfied). Each time the populations remain unchanged. Now try varying N_{10} with $N_{10} = 6$. This time, the populations will have to differ. However, by the half-way point of the time interval, they are essentially identical again. The same occurs when varying N_{20}.

Finally, the parameter T even exhibits this behavior, in a way. Vary with $T = 10$ using (22.2). During the interval $0 \leq t \leq 5$, the controls are vastly different, yet the populations remain identical. During the interval $5 \leq t \leq 10$, the second system becomes highly dynamic. The second pesticide regimen is lower on average than the first, because it is forced to meet the same isoperimetric constraint over a longer time interval. This leads to a gradual, then sudden rise in the pest population in the second half. A higher dose of pesticide near the end, along with a resurgent predator population, brings the pest population back down. However, it is still much higher than the final pest population of the first system. Also, in this simulation, the predator population does not fade out, but ends higher than it began.

Chapter 23

Discrete Time Models

For many populations, births and growth occur in regular (predictable) times each year (or each cycle). Discrete time models are well suited to describe the life histories of organisms with discrete reproduction and/or growth. For example, the Beverton-Holt stock-recruitment model for a population N_t at time t is

$$N_{t+1} = rN_t \left(1 + N_t \frac{r-1}{K}\right)^{-1}.$$

Kot [107] gives many examples of discrete time equations modeling plants, insects, fish, birds, and mammals. For systems of discrete time models and many applications, see Caswell's book on matrix models [28].

The theory of optimal control can be adapted to discrete time models. Namely, instead of continuous time ODEs as the underlying dynamics, as has been seen so far, we can instead consider state systems given by discrete difference equations. In this chapter, we give a brief derivation of the necessary conditions that the optimal control, state, and adjoint must satisfy in the case of one state and one control (without bounds). We also discuss multiple states and/or controls. The reader will immediately notice the similarity with the continuous time problems. All other problem types (states fixed at both endpoints, control bounds, etc.) follow by more or less the same techniques. For necessary conditions with more generality, see the work by Halkin [78]. For more information and examples of optimal control of discrete time models, see [33, 169].

23.1 Necessary Conditions

We will use subscripts to denote our time step throughout this chapter. Given a control $u = (u_0, u_1, \ldots, u_{T-1})$ and an initial state x_0, the corresponding state equation is given by the difference equation

$$x_{k+1} = g(x_k, u_k, k)$$

for $k = 0, 1, 2, \ldots, T-1$. Note that the state has one more component than the control

$$x = (x_0, x_1, \ldots, x_T),$$

where the first state component x_0 is given. The given first state component x_0 is analogous to the initial condition in continuous optimal control problems.

We define the objective functional as

$$J(u) = \phi(x_T) + \sum_{k=0}^{T-1} f(x_k, u_k, k)$$

where we include a payoff term. We seek to maximize $J(u)$ over vectors u in $\mathbb{R}^T$. We assume that f and g are continuously differentiable functions of their arguments.

Suppose $u^* = (u_0^*, u_1^*, \ldots, u_{T-1}^*)$ achieves the (finite) maximum of $J(u)$ over all vectors $u \in \mathbb{R}^T$, and let $x^* = (x_0, x_1^*, \ldots, x_T^*)$ be the corresponding state. Let $u^\epsilon = u^* + \epsilon h$ be another control, where h is any T-dimensional vector and $\epsilon \in \mathbb{R}$. Let x^ϵ be the corresponding state. We differentiate J with respect to u at u^* in the h direction,

$$\begin{aligned}
0 &= \lim_{\epsilon \to 0} \frac{J(u^* + \epsilon h) - J(u)}{\epsilon} \\
&= \sum_{k=0}^{T-1} \left(\frac{\partial f}{\partial x_k}(x_k^*, u_k^*, k)\psi_k + \frac{\partial f}{\partial u_k}(x_k^*, u_k^*, k)h_k \right) + \phi'(x_T^*)\psi_T
\end{aligned} \tag{23.1}$$

where the "sensitivity" $\psi = (\psi_0, \psi_1, \ldots, \psi_T)$ is defined by

$$\psi = \left.\frac{dx^\epsilon}{d\epsilon}\right|_{\epsilon=0} = \lim_{\epsilon \to 0} \frac{x^\epsilon - x^*}{\epsilon}.$$

The sensitivity is the analogue of $\frac{dx^\epsilon}{d\epsilon}$ at $\epsilon = 0$ from Chapter 1. One can think of the sensitivity ψ as the directional derivative of the control-to-state map $u \to x$ at u^* in the direction h. The existence of such a limit of those difference quotients follows from the regularity properties of f and g (as in Chapter 1). By differentiating and using the Chain Rule, we can show that ψ satisfies the *sensitivity* equations:

$$\psi_0 = 0,$$
$$\psi_{k+1} = \frac{\partial g}{\partial x_k}(x_k^*, u_k^*, k)\psi_k + \frac{\partial g}{\partial u_k}(x_k^*, u_k^*, k)h_k \quad \text{for } k = 0, 1, \ldots, T-1.$$

We choose our adjoint λ to satisfy the system

$$\lambda_k = \frac{\partial g}{\partial x_k}(x_k^*, u_k^*, k)\lambda_{k+1} + \frac{\partial f}{\partial x_k}(x_k^*, u_k^*, k) \quad \text{for } k = 0, 1, \ldots, T-1,$$
$$\lambda_T = \phi'(x_T^*) \quad \text{(transversality condition)}.$$

Substituting out the $\dfrac{\partial f}{\partial x_k}$ terms from our adjoint system, equation (23.1) becomes

$$\begin{aligned}
0 &= \sum_{k=0}^{T-1}\left[\psi_k\left(\lambda_k - \frac{\partial g}{\partial x_k}(x_k^*, u_k^*, k)\lambda_{k+1}\right) + \frac{\partial f}{\partial u_k}(x_k^*, u_k^*, k)h_k\right] \\
&\quad + \phi'(x_T^*)\psi_T. \\
&= \sum_{k=0}^{T-1}\lambda_k\psi_k + \sum_{k=0}^{T-1}\left[\frac{\partial f}{\partial u_k}(x_k^*, u_k^*, k)h_k - \psi_k\frac{\partial g}{\partial x_k}(x_k^*, u_k^*, k)\lambda_{k+1}\right] \\
&\quad + \phi'(x_T^*)\psi_T.
\end{aligned} \quad (23.2)$$

Using $\psi_0 = 0$, and then a change of index on the first sum, we transform equation (23.2) to

$$\begin{aligned}
0 &= \sum_{k=1}^{T-1}\lambda_k\psi_k + \sum_{k=0}^{T-1}\left[\frac{\partial f}{\partial u_k}(x_k^*, u_k^*, k)h_k - \psi_k\frac{\partial g}{\partial x_k}(x_k^*, u_k^*, k)\lambda_{k+1}\right] + \phi'(x_T^*)\psi_T \\
&= \sum_{k=0}^{T-2}\psi_{k+1}\lambda_{k+1} + \sum_{k=0}^{T-1}\left[\frac{\partial f}{\partial u_k}(x_k^*, u_k^*, k)h_k - \psi_k\frac{\partial g}{\partial x_k}(x_k^*, u_k^*, k)\lambda_{k+1}\right] \\
&\quad + \phi'(x_T^*)\psi_T \\
&= \sum_{k=0}^{T-2}\lambda_{k+1}\left(\psi_{k+1} - \frac{\partial g}{\partial x_k}(x_k^*, u_k^*, k)\psi_k\right) + \sum_{k=0}^{T-1}\frac{\partial f}{\partial u_k}(x_k^*, u_k^*, k)h_k \\
&\quad - \lambda_T\frac{\partial g}{\partial x_{T-1}}(x_{T-1}^*, u_{T-1}^*, T-1)\psi_{T-1} + \phi'(x_T^*)\psi_T.
\end{aligned}$$

Making substitutions via the sensitivity equations, we obtain

$$0 = \sum_{k=0}^{T-2} \lambda_{k+1} \frac{\partial g}{\partial u_k}(x_k^*, u_k^*, k) h_k + \sum_{k=0}^{T-1} \frac{\partial f}{\partial u_k}(x_k^*, u_k^*, k) h_k$$

$$- \lambda_T \left(\psi_T - \frac{\partial g}{\partial u_{T-1}}(x_{T-1}^*, u_{T-1}^*, T-1) h_{T-1} \right) + \phi'(x_T^*) \psi_T$$

$$= \sum_{k=0}^{T-1} \lambda_{k+1} \frac{\partial g}{\partial u_k}(x_k^*, u_k^*, k) h_k + \sum_{k=0}^{T-1} \frac{\partial f}{\partial u_k}(x_k^*, u_k^*, k) h_k - \lambda_T \psi_T + \phi'(x_T^*) \psi_T$$

$$= \sum_{k=0}^{T-1} h_k \left(\lambda_{k+1} \frac{\partial g}{\partial u_k}(x_k^*, u_k^*, k) + \frac{\partial f}{\partial u_k}(x_k^*, u_k^*, k) \right),$$

where the adjoint transversality condition is used to gain the last equality (ψ_T terms drop out). This equality holds for any vector h, which implies the optimality condition

$$\lambda_{k+1} \frac{\partial g}{\partial u_k}(x_k^*, u_k^*, k) + \frac{\partial f}{\partial u_k}(x_k^*, u_k^*, k) = 0 \quad \text{for } k = 0, 1, \ldots, T-1.$$

To express these conditions in terms of a Hamiltonian, define the Hamiltonian at time k by

$$H_k = f(x_k, u_k, k) + \lambda_{k+1} g(x_k, u_k, k), \quad \text{for } k = 0, 1, \ldots, T-1.$$

The necessary conditions become

$$\lambda_k = \frac{\partial H_k}{\partial x_k}$$

$$\lambda_T = \phi'(x_T^*)$$

$$\frac{\partial H_k}{\partial u_k} = 0 \text{ at } u^*.$$

Note the similarity between these conditions and the necessary conditions derived in Chapter 3. Here, the adjoint equation is missing its minus sign and is a (backwards) difference equation for λ_k, not a differential equation. The Hamiltonian here is defined in a subtly different way. The adjoint which appears in Hamiltonian at time k, H_k, is indexed forward one step in time. We should mention some authors choose to index their adjoint variables differently than is done here (e.g., [33]). This leads to a Hamiltonian which is completely analogous to the continuous time problems, but alters the necessary conditions slightly from those given here. Now we focus on a simple example.

Example 23.1

$$\min_u \sum_{k=0}^{2} \frac{1}{2}[x_k^2 + Bu_k^2]$$

subject to $x_{k+1} = x_k + u_k$ for $k = 0, 1, 2,$

$x_0 = 5.$

The control is the source or sink for the state. Since we are seeking to minimize the states and the cost of control input (with squared terms), we would expect the optimal control components to be negative in order to bring the state components down.

We have $T = 3$ steps here and take $B = 1$ for this simple calculation. Note that in our objective functional, there is no $\phi(x_T)$ term, meaning the problem does not depend on the state x_T at the last step. Our k^{th} Hamiltonian is:

$$H_k = \frac{1}{2}[x_k^2 + u_k^2] + \lambda_{k+1}(x_k + u_k).$$

Our necessary conditions are

$$\lambda_k = \frac{\partial H_k}{\partial x_k} = x_k + \lambda_{k+1} \quad \text{for } k = 0, 1, 2,$$

$$\lambda_3 = 0,$$

$$0 = \frac{\partial H_k}{\partial u_k} = u_k + \lambda_{k+1} \quad \text{at } u_k^*.$$

Substituting the characterization of the optimal control into the state equation

$$x_{k+1} = x_k - \lambda_{k+1} \quad \text{for } k = 0, 1, 2.$$

Since these conditions are very simple, we can now solve for the adjoints and state values explicitly. Using the new state equation above and the adjoint difference equation, and starting with $\lambda_3 = 0$,

$$x_3 = x_2 \quad \text{and} \quad \lambda_2 = x_2.$$

Continuing,

$$x_2 = x_1 - \lambda_2$$

implies

$$x_2 = x_1 - x_2 \Rightarrow x_2 = \frac{x_1}{2}.$$

Then the adjoint difference equation, $\lambda_1 = x_1 + \lambda_2 = x_1 + x_2$, gives

$$x_1 = x_0 - \lambda_1 = x_0 - (x_1 + x_2) = 5 - \frac{3x_1}{2}.$$

This is easily solved to give $x_1^* = 2$. Substituting back yields $x_2^* = 1$, $x_3^* = 1$. Using the state equation, our optimal control values are $u_0^* = -3$, $u_1^* = -1$, $u_2^* = 0$.

If we take the number of steps to be much larger, then one would choose to solve this problem numerically. The standard forward-backward sweep method developed in Chapter 4 is still applicable in the discrete time case, and, in fact, is even easier to implement. Indeed, recall that in the previous labs, during the forward and backward sweeps, we estimated the differential equations by discretizing the time interval and employing a Runge-Kutta method. In the discrete time case, there is no need for such an estimation, since we can use the difference equations directly. The rest of the sweep method works as before. However, discrete problems sometimes have convergence problems when the time step is not small. In this simple example, we took the time step to be 1. Sometimes the size of the balancing constant B can be adjusted to achieve convergence. Figure 23.1 gives the results when $T = 20$, $B = 100$, and $x_0 = 5$.

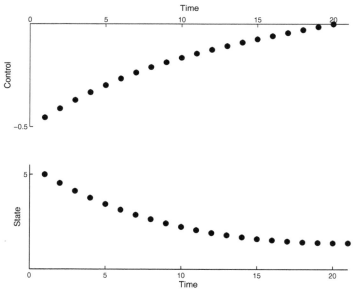

FIGURE 23.1: Optimal control and state for Example 23.1 with 20 time steps.

23.2 Systems Case

If one has multiple states and/or controls, the procedure is analogous to the systems case for differential equations. Namely, there is an adjoint variable for each state variable.

Given controls $u_i = (u_{i,0}, u_{i,1}, ..., u_{i,(T-1)})$, for $i = 1, 2, \ldots, m$, where the second subscript (in the components of the vector) denotes the time step, the corresponding state equations are given by

$$x_{j,k+1} = g_j(x_{1,k}, \ldots, x_{n,k}, u_{1,k}, \ldots, u_{m,k}, k)$$

for $k = 0, 1, 2, \ldots, T-1, j = 1, 2, \ldots, n$. Note that each state has one more component than the controls:

$$x_j = (x_{j,0}, x_{j,1}, \ldots, x_{j,T}).$$

In this notation, there are m controls, n states, and T time steps. Define the objective functional as

$$J(u) = \phi(x_{1,T}, \ldots, x_{n,T}) + \sum_{k=0}^{T-1} f(x_{1,k}, \ldots, x_{n,k}, u_{1,k}, \ldots, u_{m,k}, k).$$

The same analysis as before yields analogous necessary conditions. Again we express them in terms of the Hamiltonian,

$$H_k = f(x_{1,k}, \ldots, x_{n,k}, u_{1,k}, \ldots, u_{m,k}, k)$$
$$+ \sum_{j=1}^{n} \lambda_{j,k+1} g_j(x_{1,k}, \ldots, x_{n,k}, u_{1,k}, \ldots, u_{m,k}, k),$$

the necessary conditions are

$$\lambda_{j,k} = \frac{\partial H_k}{\partial x_{j,k}}$$

$$\lambda_{j,T} = \frac{\partial \phi}{\partial x_{j,T}}(x_{1,T}, \ldots, x_{n,T})$$

$$\frac{\partial H_k}{\partial u_{i,k}} = 0 \text{ at } (u_{1,k}^*, \ldots, u_{m,k}^*)$$

for all $k = 0, 1, \ldots, T-1$, $j = 0, 1, \ldots, n$, and $i = 0, 1, \ldots, m$.

Note that a system of discrete time models can be used to express spatial structure. The different state variables may represent populations or resources in different geographic areas; see the book by Tan and Bennett. [174]

Example 23.2 Consider this simple system with 2 states, 1 control, and 3 time steps.

$$\max_u x_{2,3} - \sum_{k=0}^{2} \left(\frac{1}{2}u_k^2 + x_{1,k}^2\right)$$

subject to $x_{1,k+1} = u_k + x_{1,k}$, $x_{1,0} = 1$,
$x_{2,k+1} = x_{1,k} + x_{2,k}$, $x_{2,0} = 0$.

The Hamiltonian here is

$$H_k = -\frac{1}{2}u_k^2 - x_{1,k}^2 + \lambda_{1,k+1}(u_k + x_{1,k}) + \lambda_{2,k+1}(x_{1,k} + x_{2,k}),$$

which gives

$$0 = \frac{\partial H_k}{\partial u_k} = -u_k + \lambda_{1,k+1} \Rightarrow u_k = \lambda_{1,k+1},$$

$$\lambda_{1,k} = \frac{\partial H_k}{\partial x_{1,k}} = -2x_{1,k} + \lambda_{1,k+1} + \lambda_{2,k+1}, \quad \lambda_{1,3} = 0,$$

$$\lambda_{2,k} = \frac{\partial H_k}{\partial x_{2,k}} = \lambda_{2,k+1}, \quad \lambda_{2,3} = 1,$$

since $\phi(x_{2,3}) = x_{2,3}$. It follows from the last equation that $\lambda_2 = (1, 1, 1, 1)$. Substituting $\lambda_{1,k+1}$ for u_k, we have

$$x_{1,k+1} = x_{1,k} + \lambda_{1,k+1},$$
$$\lambda_{1,k} = -2x_{1,k} + \lambda_{1,k+1} + 1.$$

Using $\lambda_{1,3} = 0$, and alternating between the two above equations, we see

$$x_{1,3} = x_{1,2} + \lambda_{1,3} = x_{1,2},$$
$$\lambda_{1,2} = -2x_{1,2} + \lambda_{1,3} + 1 = -2x_{1,2} + 1,$$
$$x_{1,2} = x_{1,1} + \lambda_{1,2} = x_{1,1} - 2x_{1,2} + 1, \Rightarrow x_{1,2} = \frac{1}{3}(x_{1,1} + 1),$$
$$\lambda_{1,1} = -2x_{1,1} + \lambda_{1,2} + 1 = -2x_{1,1} - 2x_{1,2} + 2 = -\frac{8}{3}x_{1,1} + \frac{4}{3},$$
$$x_{1,1} = x_{1,0} + \lambda_{1,1} = 1 - \frac{8}{3}x_{1,1} + \frac{4}{3} \Rightarrow x_{1,1} = \frac{7}{11}.$$

Substituting back, $x_1^* = (1, 7/11, 6/11, 6/11)$ and $x_2^* = (0, 1, 18/11, 24/11)$. From the difference equation for x_1, we have $u = (-4/11, -1/11, 0)$.

Discrete Time Models

Example 23.3 Invasive plant populations frequently consist of a large main focus and several smaller outlier populations. Management of such an invasive population requires decisions about whether to control the main focus and/or the outlier populations and about the timing of control actions.

Consider an invasion of a plant species that consists of a finite number of satellites with a main focus. Moody and Mack [145] look at two control scenarios for this problem. Namely, allow either all satellites to be eradicated or an amount of the main focus to be removed, that amount being equivalent to the total area of all foci (assuming the focus and the satellites were circles). They then allowed the invasive populations to expand and found that the total area invaded was always greater, after a sufficiently long period of time, for the second scenario (removing area from the main focus). This is not surprising as after an initial transient period, all circles become sufficiently large that asymptotically the increase in area becomes the same. Thus having more satellites remaining will give greater area than just one remaining.

Recently a model by Whittle et. al. [179] generalized this work to include a finite time horizon and a finite number of steps at which controls can be applied. The state equations for the radius of the invasion over time are given by

$$r_{j,t+1} = \left(r_{j,t} + \frac{r_{j,t}k}{\varepsilon + r_{j,t}}\right)(1 - u_{j,t})$$

for $t = 0, 1, \ldots, T-1$, where $r_{j,t}$ represents the radius of the focus j at time t. The initial size of the foci are given and will determine which is the main focus. The spread rate is given by k and is scaled by $\frac{r}{\varepsilon+r}$ for ε small. The scaling ensures that if a focus is eradicated, $r_{j,t} = 0$, it remains eradicated and does not grow back. The control coefficient $u_{j,t}$ is the amount of radius decrease due to control at time t for the focus j. Note that in discrete time models, the order of the events is crucial; here the growth happens before the removal by the control. We want to minimize the area covered by the invasive species at the final time and the cost of the control over the whole time period. With controls $u = (u_1, u_2, \ldots, u_n)$ where $u_j = (u_{j,0}, u_{j,1}, \ldots, u_{j,T-1})$, the objective functional is given by

$$J(u) = \sum_{j=1}^{n}\left[r_{j,T}^2 + B\sum_{t=0}^{T-1} u_{j,t}^2\right]$$

where B is a positive-valued weight parameter. We seek to minimize $J(u)$ over controls with components $0 \leq u_{j,t} \leq 1$.

The Hamiltonian is given by

$$H_t = \sum_{j=1}^{n}\left[Bu_{j,t}^2 + \lambda_{j,t+1}\left[\left(r_{j,t} + \frac{r_{j,t}k}{\varepsilon + r_{j,t}}\right)(1 - u_{j,t})\right]\right]. \qquad (23.3)$$

Using the necessary conditions, we have

$$\lambda_{j,t} = \frac{\partial H_t}{\partial r_{j,t}} = \lambda_{j,t+1}(1 - u_{j,t})\left(1 + \frac{\varepsilon k}{(\varepsilon + r_{j,t})^2}\right)$$

for $t = 1, 2, \ldots, T - 1$, with the transversality conditions

$$\lambda_{j,T} = 2r_{j,T}.$$

Furthermore the optimal control is given by

$$u_{j,t}^* = \max\left\{0, \min\left\{\frac{\lambda_{j,t+1}}{2B}\left(r_{j,t} + \frac{r_{j,t}k}{\varepsilon + r_{j,t}}\right), 1\right\}\right\}.$$

This characterization is obtained from solving the optimality condition

$$\frac{\partial H}{\partial u_{j,t}} = 2Bu_{j,t}^* - \lambda_{j,t+1}\left(r_{j,t} + \frac{r_{j,t}k}{\varepsilon + r_{j,t}}\right).$$

Although we have not discussed bounds in the discrete time case, the necessary conditions follow in exactly the same way as before. In particular, the Hamiltonian is still maximized (or minimized) pointwise by the optimal control. We will see numerical results for a similar example in the next lab.

For an optimal control example of a discrete model for pest control, specifically for gypsy moths, see [178]. For a review of "biocontrol" models of pests and invasives, see [81].

23.3 Exercises

Exercise 23.1 (from [126]) Solve

$$\min_{u} \sum_{k=0}^{4} u_k^2$$

$$\begin{aligned}
\text{subject to} \quad & x_{k+1} = 2x_k + u_k, \quad \text{for } k = 0, 1, \ldots, 4, \\
& x_0 = 3, \; x_5 = 0, \\
& 0 \leq u_k \leq 1 \quad \text{for } k = 0, 1, \ldots, 4.
\end{aligned}$$

Exercise 23.2 (from [169]) The following problem has a singular case at one time step. Solve

$$\min_u \sum_{k=0}^{5} \frac{1}{2}[x_k^2]$$

subject to $x_{k+1} = x_k + u_k$ for $k = 0, 1, \ldots, 5$,

$x_0 = 5$,

$-1 \leq u_k \leq 1$ for $k = 0, 1, \ldots, 5$.

Exercise 23.3 State the adjoint equations and optimality condition for this system problem, which is a simple pest control problem. View the y population as the pest to be controlled with u.

$$\max_u \sum_{k=0}^{T} \frac{1}{2}[x_k - u_k^2]$$

subject to $x_{k+1} = x_k + x_k(1 - x_k) - x_k y_k$ for $k = 0, 1, \ldots, T$,

$y_{k+1} = y_k + x_k y_k - u_k y_k$ for $k = 0, 1, \ldots, T$,

$x_0 = 5$, $y_0 = 10$,

$0 \leq u_k \leq 1$ for $k = 0, 1, \ldots, T$.

Chapter 24

Lab 14: Invasive Plant Species

Reconsider Example 23.3, where an invasive plant species was modeled. The species consists of a finite number of satellites with a main focus. We wish to minimize the total area covered by the invader at the end of a finite time period. A discrete time model is used. The optimal control problem we considered before was

$$\min_u \sum_{j=1}^{n} \left[r_{j,T}^2 + B \sum_{t=0}^{T-1} u_{j,t}^2 \right]$$

$$\text{subject to } r_{j,t+1} = \left(r_{j,t} + \frac{r_{j,t} k}{\varepsilon + r_{j,t}} \right)(1 - u_{j,t}), \; r_{j,0} = \rho_j, \quad (24.1)$$

$$0 \le u_{j,t} \le 1 \text{ for } j = 0, 1, \ldots, T-1,$$

where B is a weight parameter, and ρ_j are the known initial conditions. Recall, the state $r_{j,t}$ is the radius length of focus j at time step t, the control $u_{j,t}$ is the amount of radius decrease for focus j at time step t, and k is the spread rate.

To simplify matters, we will fix some of the constants for this lab. The number of foci n will be set to five, so that there is one main focus and four satellites. The initial values will be set to $\rho_1 = 0.5$, $\rho_2 = 1$, $\rho_3 = 1.5$, $\rho_4 = 2$, $\rho_5 = 10$, so that the main focus is denoted by $j = 5$. Further, the arbitrary constant ε will be fixed at $\varepsilon = 0.01$.

Further, we wish to avoid the transversality condition $\lambda_{j,T} = 2r_{j,T}$. This type of boundary condition cannot be handled by the standard forward-backward sweep (the adapted sweep can be used, of course; see Exercise 24.1). Therefore, we change to

$$\phi(x_{1,T}, \ldots, x_{5,T}) = \sum_{j=1}^{5} x_{j,T},$$

which gives the transversality conditions $\lambda_{j,T} = 1$. Therefore, our new optimal control problem is

$$\min_u \sum_{j=1}^{5} \left[r_{j,T} + B \sum_{t=0}^{T-1} u_{j,t}^2 \right]$$

subject to $r_{j,t+1} = \left(r_{j,t} + \dfrac{r_{j,t}k}{\varepsilon + r_{j,t}}\right)(1 - u_{j,t}), \quad r_{j,0} = \rho_j,$ (24.2)

$0 \leq u_{j,t} \leq 1.$

To be clear, we are not claiming (24.1) and (24.2) are *equivalent* optimal control problems. They will no doubt produce different results. However, by only removing the square, we have not drastically altered the dynamics of the problem. This new problem should have solutions of the same flavor as the original.

Before beginning, we discuss the slight changes in the code, as compared to previous forward-backward sweep routines. The first thing to note, is that while our vectors have been indexed from 0 to T, MATLAB indexes vectors starting with 1. Therefore, the states and adjoints, which should run from 0 to T, actually start at 1 and end with $T+1$. Similarly, the controls should end with $T-1$, but end with T. This is only a superficial difference, occurring entirely in the code. In fact, the t variable, as defined, gives the correct index values. Namely, $t(1) = 0, t(2) = 1, \ldots, t(T+1) = T$. This t will be used to graph the solutions correctly indexed.

The other minor changes involve the differing vector lengths. Before, we updated the characterization of the control all at once, via vectors. Here, we must be more careful, as the controls u_j have one less term than the states and adjoints. Therefore, it is done term-by-term in a *for* loop. Then, the standard convex combination is used. Second, instead of outputting the solutions as the dummy matrix y, we must use two matrices (y and z). The y variable contains the states (running from 0 to T), and the controls (running from 0 to $T-1$) are given by z.

Everything else in the code should be familiar. As we noted in the previous chapter, instead of a Runge-Kutta 4 routine, the difference equations are solved exactly as stated.

In MATLAB run the program *lab14*. To begin, try the values

$$\boxed{B = 1 \quad k = 1 \quad T = 10}, \qquad (24.3)$$

and do not vary any parameters. The optimal solutions are displayed in a different manner than normal. All five state variables are plotted together on the left. For emphasis, the state variable representing the main foci, r_5, is plotted in red, with the values denoted by circles. The remaining four states are in blue, marked by x's. Also, while the x's and circles denote the values of the states, dotted lines connect the points. This is done in order to make the plots easier to see and read. The five controls are plotted together on the right, with the same color and marker scheme.

This first simulation illustrates one of the major differences between continuous time and discrete time models. Here, choosing the control to be 1 at any time immediately forces the corresponding radius to 0. The optimal strategy depicted is to use no control for the first 8 years, then maximum control in the last year, for all radii. This causes the radii to grow naturally for the majority of the time interval, only to be completely eradicated in the last year. We note, with these parameters, any controls with value 1 for one year, and 0 at all others, would produce the same final state values, and are therefore optimal as well.

Enter (24.3), varying with $k = 0.5$. The radii for the first system (with $k = 1$) are depicted as before. The main focus for the second system (with $k = 0.5$) is plotted in purple, with circle markers. The remaining four foci are plotted in green with x markers. The controls are similar. The control strategy is the same for both simulations. Namely, use no control until the last year, then use maximum control. The foci grow naturally until the last year, when they are pushed to 0. The only difference occurs during the natural growth phase, where the result of the different k values is clear.

This control strategy, that of maximum control at the end, is optimal in these cases primarily because of the weight parameter $B = 1$. Minimizing the final radii values and the total control are of equal importance, which allows this harsh control strategy. Suppose we increase the importance of keeping the control low. Enter (24.3), varying with $B = 2$. The optimal solutions are identical. We still are neglecting the control too much. Vary with $B = 10$. This causes a change in strategy. The second system employs controls with less total square sum. As such, the radii do not decrease all the way to 0. Note, all controls in the second simulation steadily increase, and the (second) main focus steadily decreases. However, the other four foci increase at first, and actually end at values higher than their initial positions. We also point out that while the five foci begin at varying values, they end at exactly the same value (0) in the first simulation, and very near the same value in the second.

Now examine

$$\boxed{B = 5 \quad k = 1 \quad T = 10}. \tag{24.4}$$

Here, the optimal strategy is a mixed one. For the four minor foci, we should use a steadily increasing control, as we just saw. On the other hand, the main focus should be eradicated via a max control at the end of the time interval. With this medium B value, it is still advantageous to eliminate the main focus, which begins much larger than the others, but to only maintain the others. With

$$\boxed{B = 4.19 \quad k = 1 \quad T = 10} \tag{24.5}$$

the main focus and the largest minor focus are eliminated, while the others are maintained. Try

$$\boxed{B = 4.1 \quad k = 1 \quad T = 10} \tag{24.6}$$

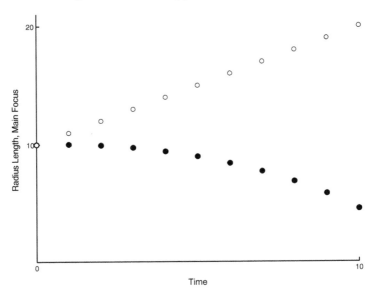

FIGURE 24.1: Optimal radius value r_5^* with $B = 10$, $k = 1$, and $T = 10$ plotted in solid circles, together with r_5 with no control applied, in open circles.

and

$$\boxed{B = 4 \quad k = 1 \quad T = 10} \tag{24.7}$$

to see the other possibilities.

Return to the variable k. Enter

$$\boxed{B = 10 \quad k = 1 \quad T = 10} \tag{24.8}$$

varying with $k = 0.5$. With less natural growth, the controls in the second simulations need not increase so dramatically at the end, but instead can be more balanced. We see the controls in the second system begin at higher values than their respective counterparts in the first system, but end at lower values. Also, the foci in the second system end with lower radii values. Now vary (24.8) with $k = 3$. The natural growth rate here is too high, and the less aggressive maintenance strategy is no longer optimal. Even with $B = 10$, the second system has returned to using eradication during the last year.

Finally, we examine the role of T. Enter (24.8), varying with $T = 20$. The new control regimes are relatively similar to their counterparts, only stretched out over 20 years instead of 10. The resulting radii are quite different for much of the interval. Note, for example, the second main focus experiences a period of increase, whereas the first only decreases. However, all radii end at nearly the same value. In this case, the optimal strategy was to alter the controls in order to achieve the same result, more or less.

Exercise 24.1 Write an adapted forward-backward sweep code for the original discrete problem

$$\min_u \sum_{j=1}^{5} \left[r_{j,T}^2 + B \sum_{t=0}^{T-1} u_{j,t}^2 \right]$$

subject to $r_{j,t+1} = \left(r_{j,t} + \dfrac{r_{j,t} k}{\varepsilon + r_{j,t}} \right) (1 - u_{j,t}), \quad r_{j,0} = \rho_j,$

$0 \leq u_{j,t} \leq 1,$

with the same values for ρ_j and ε. You should also be able to write a code where ρ_j and ε are inputs. Readers with more MATLAB experience can write a code where n is an input.

Chapter 25

Partial Differential Equation Models

There are many population models that involve a spatial component [27, 45, 155]. For example, recall the bioreactor model used in Example 12.4 and Lab 12. There, we assumed contaminant and bacteria levels were spatially uniform, but this may not always be a valid assumption. Different locations in the bioreactor may promote or discourage bacterial growth. In this case, we would add a spatial variable (or variables). The second volume of Murray [150] contains many different examples of models with spatial features. Of course, depending on the scale of the spatial resolution, the introduction of space variables can alter our models from ODEs (with just time as the underlying variable) to partial differential equations (PDEs). If the spatial structure gives a metapopulation model of ODEs [79], then the systems approach to optimal control already presented is appropriate. We now turn our attention to consideration of optimal control of PDEs.

J.-L. Lions laid the foundation of the basic ideas of optimal control of partial differential equations in the 1970's [129]. There is no complete generalization of Pontryagin's Maximum Principle to partial differential equations, but the book by Li and Yong [128] deals with corresponding "maximum principle" type results. There are also some counterexamples for certain infinite dimensional systems (systems of PDEs are considered infinite dimensional systems, but ODEs are finite dimensional). The examples we treat here have maximum principle type results. We also call the reader's attention to the books by Barbu, Lasiecka and Triggiani, Fattorini, and Mordukhovich for a variety of results on optimal control of PDEs [5, 10, 11, 57, 110, 111, 112, 147].

Choosing the underlying solution space for the states is a crucial feature for optimal control of PDEs. Classical solutions (solutions with all the derivatives occurring in the PDE being continuous) will not exist for most nonlinear PDE problems. Deciding in what "weak" sense we are solving the PDEs is essential. We refer to Evans [56] and Friedman [66] for the rigorous definitions of Sobolev spaces and weak derivatives and give only an informal treatment. This chapter will require more background in analysis and PDEs than the other chapters.

Let Ω be an open, connected subset of $\mathbb{R}^n$. From now on, x (and occasionally y) will be the space variable associated to Ω. One can think of a weak derivative as the function which makes the appropriate integration by parts work: for u and v, which are integrable (in the Lebesgue sense) on Ω, we say v is the weak x_i-derivative of u if

$$\int_\Omega u\,\phi_{x_i}\,dx = -\int_\Omega v\,\phi\,dx$$

for all ϕ in $C_c^\infty(\Omega)$, which is the set of all infinitely differentiable functions on Ω with compact support.

For most parabolic PDE control problems, such as those involving diffusion, the appropriate solution space is $L^2([0,T]; H_0^1(\Omega))$. Roughly speaking, this space consists of all functions square-integrable in time with two weak derivatives in space, which are also square-integrable. The control set is frequently the Lebesgue integrable functions, which have specified upper and lower bounds.

The general idea of optimal control of PDEs starts with a PDE with state solution w and control u. Take A to be a partial differential operator with appropriate initial conditions (IC) and boundary conditions (BC),

$$Aw = f(w,u) \text{ in } \Omega \times [0,T], \text{ along with } BC, IC,$$

assuming the underlying variables are x for space and t for time. We are treating problems with space and time variables, but one could treat steady state problems with only spatial variables [26, 122, 125].

Again, the objective functional represents the goal of the problem; here we write our functional in an integral form. We seek to find the optimal control u^* in an appropriate control set such that

$$J(u^*) = \inf_u J(u),$$

with objective functional

$$J(u) = \int_0^T \int_\Omega g(x,t,w(x,t),u(x,t))\,dx\,dt.$$

After specifying a control set and a solution space for the states, one can usually obtain the existence of a state solution given a control. Namely, for a given control u, there exists a state solution $w = w(u)$, showing the dependence of w on u.

25.1 Existence of an Optimal Control

Proving the existence of an optimal control in the PDE case requires more work than in the ODE case. A *priori* estimates of the norms of the states in the solution space are needed to justify convergence. If the controls are bounded above and below, one can usually obtain corresponding bounds in the solution space for the states. This boundedness gives the existence of a minimizing sequence u_n where

$$\lim_{n \to \infty} J(u_n) = \inf_u J(u).$$

In the appropriate weak sense, this usually gives

$$u_n \rightharpoonup u^* \quad \text{in} \quad L^2(\Omega \times [0, T]),$$
$$w_n = w(u_n) \rightharpoonup w^* \quad \text{in the solution space,}$$

for some u^* and w^*. One must show $w^* = w(u^*)$, which means that the solution depends continuously on the controls. We must also show that u^* is an optimal control, i.e.,

$$J(u^*) = \inf_u J(u).$$

To derive the necessary conditions like we did in Chapter 1, we need to differentiate the objective functional with respect to the control, namely, differentiate the map

$$u \longmapsto J(u).$$

Note that $w = w(u)$ usually contributes to $J(u)$, so we must also differentiate the map

$$u \longmapsto w(u).$$

In the usual well-posed PDE problem, continuous dependence of the solution on the control would imply continuity of this map, but differentiable dependence is needed here.

25.2 Sensitivities and Necessary Conditions

The map $u \mapsto w(u)$ is weakly differentiable in the directional derivative sense (Gateaux):

$$\lim_{\epsilon \to 0} \frac{w(u + \epsilon l) - w(u)}{\epsilon} = \psi.$$

The function ψ is called the *sensitivity* of the state with respect to the control. It is analogous to the

$$\left. \frac{\partial x^\epsilon}{\partial \epsilon} \right|_{\epsilon=0}$$

term from the proof of the necessary conditions in Chapter 1. In the ODE case, the adjoint is chosen to make the sensitivity terms drop out, but here we need the sensitivity PDE to find the adjoint PDE. *A priori* estimates of the difference quotients in the norm of the solution space give the existence of the limit function ψ and that it solves a PDE, which is linearized version of the state PDE

$$L\psi = F(w, l, u) \text{ with appropriate } BC, IC.$$

Note the linear operator L comes from linearizing the state PDE operator A. Usually, ψ will have zero BC and IC since $w(u + \epsilon l)$ and $w(u)$ have the same BC and IC. Differentiating the objective functional $J(u)$ with respect to u at u^*,

$$0 \leq \lim_{\epsilon \to 0^+} \frac{J(u^* + \epsilon l) - J(u^*)}{\epsilon}.$$

We use the adjoint problem λ and ψ to simplify and obtain the explicit characterization

$$u^* = G(w^*, \lambda)$$

of the optimal control. The use of λ and ψ to simplify the difference quotient of J and to derive the optimal control characterization will be clarified in the following examples. The operator in the adjoint equation is the adjoint operator (in the functional analysis sense [38]) of the operator acting on ψ in the sensitivity equation. The boundary conditions from the adjoint system come from the boundary conditions for the sensitivity equations and properties of the adjoint operator. Formally, the operator L and the adjoint operator L^* are related by

$$\langle \lambda, L\psi \rangle = \langle L^*\lambda, \psi \rangle$$

where $\langle \cdot, \cdot \rangle$ is the L^2-inner product. A key tool is integration by parts in multidimensions to throw the derivatives on ψ from the operator L onto the derivatives of λ in the operator L^*.

In a time dependent problem, the adjoint problem usually has final time conditions, like the transversality conditions that we have treated so far. The nonhomogeneous term of the adjoint equation comes from differentiating the integrand of the objective functional with respect to the state. Informally,

$$L^*\lambda = \frac{\partial(\text{integrand of } J)}{\partial w},$$

where L is the operator from the ψ PDE. We will have a characterization of an optimal control in terms of the solutions of the state and adjoint equations. This system, the state and adjoint equations together with the control characterization, is called the *optimality system*.

25.3 Uniqueness of the Optimal Control

Uniqueness of solutions to the optimality system,

$$Aw^* = f(w^*, G(w^*, \lambda)),$$
$$L^*\lambda = \frac{\partial(\text{integrand of } J)}{\partial u},$$

BC, state IC, with adjoint final time conditions,

will imply uniqueness of the optimal control, since an optimal control and corresponding state and adjoint satisfy this system. In the usual time dependent case, like a diffusion equation, the adjoint equation has final time conditions while the state equation has an initial time condition. This means that the state equation and the adjoint equation have opposite time orientations

$$w = w_0 \quad \text{for } t = 0,$$
$$\lambda = 0 \quad \text{for } t = T.$$

Typically, one can only prove uniqueness for small T [59, 117]. Numerical algorithms are usually able to solve the optimality systems for larger times. Note that an alternative approach to prove uniqueness of the optimal control is to verify directly the strict convexity of the map $u \to J(u)$; this involves calculation of the second derivative of the control-to-state map $u \to w(u)$ [118].

25.4 Numerical Solutions

In this chapter, we will only discuss briefly one numerical method, which is the PDE analogue of the methods used here on ODEs. A "forward-backward sweep" iteration method can be used to solve such optimality systems. Each sweep must be done by some type of PDE solver, like finite difference or finite elements. We repeat the iterations until convergence of successive controls and state solutions are close together. If there is a problem with uniqueness of the solutions of the optimality system, one might choose to also calculate the change in the objective functional. See books by Ahmed, Teo, Neittaanamaki, Tiba, and Troelsch [2, 152, 176] for numerical methods for optimal control problems in PDEs.

In the following examples, we will concentrate on doing the calculations of the sensitivity equation, adjoint equation, and the characterization of the optimal control. We do not treat the details of proving existence and uniqueness of the optimal control here. We will also assume the appropriate difference quotients converge to the sensitivity function, which would need to be proven in a fully justified solution. We want to emphasize the importance of doing the analysis estimates, but we feel that the background necessary is beyond the scope of this book [56]. We refer the reader to the following references [59, 117, 119] to see examples with such details. All of the following examples are parabolic PDEs because of their relevance to biological situation. One can treat elliptic PDEs for the steady state situations, but there are no initial or final time conditions involved [122]. Further, just proving that solutions of the state equation (or system) in the elliptic case exist can be difficult. See the chapter on control of PDEs in Knowles [106].

25.5 Harvesting Example

Consider the problem of harvesting in a diffusing population

$$w_t(x,t) - \alpha \Delta w(x,t) = w(x,t)(1 - w(x,t)) - u(x,t)w(x,t) \text{ in } \Omega \times (0,T),$$
$$w(x,t) = 0 \text{ on } \partial\Omega \times (0,T) \quad \text{(side boundary)},$$
$$w(x,0) = w_0(x) \geq 0 \text{ on } \Omega, t = 0 \quad \text{(bottom boundary)},$$

where $\partial\Omega$ is the boundary of Ω. The symbol Δ represents the Laplacian. In two dimensions, $\Delta w = w_{x_1 x_1} + w_{x_2 x_2}$, where $x = (x_1, x_2)$. The state $w(x,t)$ is the density of the population and the harvesting control is $u(x,t)$. Note the state equation has logistic growth $w(1-w)$ and constant diffusion coefficient α. The "profit" objective functional is

$$J(u) = \int_0^T \int_\Omega e^{-\delta t}(pu(x,t)w(x,t) - Bu(x,t)^2)\, dx\, dt,$$

which is a discounted "revenue less cost" stream. With p representing the price of harvesting population, puw represents the revenue from the harvested amount uw. We use a quadratic cost for the harvesting effort with a weight coefficient B. At first, we consider the case of a positive constant B. The coefficient $e^{-\delta t}$ is a discount term with $0 \leq \delta < 1$. For convenience, we now take the price to be $p = 1$.

A main point of interest in a fishery application is where the "marine reserves" should be placed, that is, the regions of no harvesting, $u^*(x,t) = 0$. We seek to find u^* such that

$$J(u^*) = \max_u J(u),$$

where the maximization is over all measurable controls with $0 \leq u(x,t) \leq M < 1$ a.e.. Under this set-up, we note that any state solution will satisfy

$$w(x,t) \geq 0 \quad \text{on} \quad \Omega \times (0,T),$$

by the Maximum Principle for parabolic equations [56].

First, we differentiate the $u \to w$ map. Given a control u, consider another control $u^\epsilon = u + \epsilon l$, where l is a variation function and $\epsilon > 0$. Let $w = w(u)$ and $w^\epsilon = w(u^\epsilon)$. The state PDEs corresponding to controls, u and u^ϵ, are

$$w_t - \alpha \Delta w = w(1-w) - uw$$
$$w_t^\epsilon - \alpha \Delta w^\epsilon = w^\epsilon(1-w^\epsilon) - u^\epsilon w^\epsilon.$$

We form the difference quotient

$$\frac{w^\epsilon - w}{\epsilon},$$

and find the corresponding PDE satifed by the difference quotients

$$\left(\frac{w^\epsilon - w}{\epsilon}\right)_t - \alpha \Delta \left(\frac{w^\epsilon - w}{\epsilon}\right) = \frac{w^\epsilon - w}{\epsilon} - \left(\frac{(w^\epsilon)^2 - w^2}{\epsilon}\right) - u\left(\frac{w^\epsilon - w}{\epsilon}\right) - lw^\epsilon.$$

Assume that as $\epsilon \to 0$, $w^\epsilon \to w$ and

$$\frac{w^\epsilon - w}{\epsilon} \to \psi.$$

As for the nonlinear term, note that

$$\frac{(w^\epsilon)^2 - w^2}{\epsilon} = (w^\epsilon + w)\frac{w^\epsilon - w}{\epsilon} \to 2w\psi.$$

The corresponding derivative quotients will converge

$$\left(\frac{w^\epsilon - w}{\epsilon}\right)_t - \Delta\left(\frac{w^\epsilon - w}{\epsilon}\right) \to \psi_t - \Delta \psi.$$

The resulting PDE for ψ is

$$\psi_t - \Delta \psi = \psi - 2w\psi - u\psi - lw \text{ on } \Omega \times (0,T),$$
$$\psi = 0 \text{ on } \partial\Omega \times (0,T),$$
$$\psi = 0 \text{ on } \{t=0\}.$$

Given an optimal control u^* and the corresponding state w^*, we rewrite the sensitivity PDE as

$$L\psi = -lw^*, \text{ where } L\psi = \psi_t - \Delta\psi - \psi + 2w^*\psi + u^*\psi.$$

Now we discuss the process of finding the adjoint equation. The basic idea of the L^* operator in the adjoint PDE is the following

$$\int_0^T \int_\Omega e^{-\delta t} \lambda L\psi \, dx \, dt = \int_0^T \int_\Omega e^{-\delta t} \psi (L^*\lambda + \delta\lambda) \, dx \, dt.$$

To see the specific terms of L^*, use integration by parts to see

$$\int_0^T \int_\Omega e^{-\delta t} \lambda \psi_t \, dx \, dt = \int_0^T \int_\Omega -e^{-\delta t}(-\delta\lambda + \lambda_t)\psi \, dx \, dt.$$

The boundary terms on $\Omega \times \{T\}$ and $\Omega \times \{0\}$ vanish due to λ and ψ being zero on the top and the bottom of our domain, respectively. The term with δ comes from the discount term in the objective functional. Next notice

$$\int_0^T \int_\Omega e^{-\delta t} \lambda \psi_{xx} dx \, dt = \int_0^T \int_\Omega (-e^{-\delta t}\lambda_x)\psi_x \, dx \, dt = \int_0^T \int_\Omega e^{-\delta t}\lambda_{xx}\psi \, dx \, dt$$

since λ and ψ are zero on $\partial\Omega \times (0, T)$. The linear terms of L go directly in L^* as the same types of terms. Our operator L^* and the adjoint PDE are

$$L^*\lambda = -\lambda_t - \Delta\lambda - \lambda + 2w^*\lambda + u^*\lambda$$
$$\text{adjoint PDE } L^*\lambda + \delta\lambda = u^* \text{ on } \Omega \times (0, T)$$
$$\lambda = 0 \text{ on } \partial\Omega \times (0, T)$$
$$\lambda = 0 \text{ on } \Omega \times \{t = T\}.$$

The nonhomogeneous term u in the adjoint PDE comes from

$$\frac{\partial(\text{integrand of J})}{\partial(\text{state})} = \frac{\partial(uw)}{\partial w} = u$$

where we use the integrand of J without the discount factor $e^{-\delta t}$, which came into play in the integration by parts above.

Next, we use the sensitivity and adjoint functions in the differentiation of the map $u \to J(u)$. At the optimal control u^*, the quotient is non-positive since $J(u^*)$ is the maximum value, i.e.,

$$0 \geq \lim_{\epsilon \to 0^+} \frac{J(u^* + \epsilon l) - J(u^*)}{\epsilon}.$$

Rewriting the adjoint equation as $L^*\lambda + \delta\lambda = u^*$, this limit simplifies to

$$0 \geq \lim_{\epsilon \to 0+} \int_0^T \int_\Omega e^{-\delta t} \frac{1}{\epsilon}((u^* + \epsilon l)w^\epsilon - u^* w^* - (B(u^* + \epsilon l)^2 - Bu^{*2})) \, dx \, dt$$

$$= \int_0^T \int_\Omega e^{-\delta t}[u^* \psi + lw^* - 2Bu^* l] \, dx \, dt$$

$$= \int_0^T \int_\Omega e^{-\delta t}(\psi(L^*\lambda + \delta \lambda) + lw^* - 2Bu^* l) \, dx \, dt$$

$$= \int_0^T \int_\Omega e^{-\delta t}(\lambda L\psi + lw^* - 2Bu^* l) \, dx \, dt$$

$$= \int_0^T \int_\Omega e^{-\delta t}(-\lambda lw^* + lw^* - 2Bu^* l) \, dx \, dt$$

$$= \int_0^T \int_\Omega e^{-\delta t} l(w^*(1 - \lambda) - 2Bu^*) \, dx \, dt,$$

where we used that the RHS of the ψ PDE is $-lw^*$.

On the set $\{(x,t) : 0 < u^*(x,t) < M\}$, the variation l can have any sign, because the optimal control can be modified a little up or down and still stay inside the bounds. Thus on this set, in the case that $B \neq 0$, the rest of the integrand must be zero, so that

$$u^* = \frac{w^*(1 - \lambda)}{2B}.$$

By taking the upper and lower bounds into account, we obtain

$$u^* = \min\left(M, \max\left(\frac{w^*(1 - \lambda)}{2B}, 0\right)\right).$$

This completes the analysis in the case of positive balancing constant B.

However, the $B = 0$ case is also important. In this case, we are maximizing profit (yield) only. When $B = 0$, the problem is linear in the control u. The argument above goes through with $B = 0$ until the end, before we solve for u^*. On the set $\{(x,t) : 0 < u^*(x,t) < M\}$, the variation l can have any sign, which corresponds to $\lambda = 1$. This case is called singular because the integrand of the objective functional drops out on this set. Suppose $\lambda = 1$ on some set of positive measure. By looking at the adjoint PDE, and noting the derivatives of λ are 0, we can solve for the state

$$w^* = \frac{1 - \delta}{2}.$$

Now use this constant for w in the state equation and solve for the optimal control

$$u^* = \frac{1 + \delta}{2}.$$

By an argument similar to the above one, we get

$$u^*(x,t) = \begin{cases} 0 & \text{if } \lambda > 1, \\ \frac{1+\delta}{2} & \text{if } \lambda = 1, \\ M & \text{if } \lambda < 1. \end{cases}$$

We illustrate one numerical example of this bang-bang case in Figure 25.1, taken from the work of Joshi et. al. [92]. The singular case does not occur in this numerical example. A forward-backward sweep iteration method was used to solve this problem. Each sweep was done by a finite difference scheme. In this example, there is a region in the spatial domain with no harvest, which would be considered a marine reserve, as in Example 17.5.

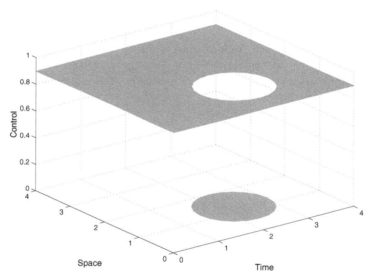

FIGURE 25.1:   Proportion to be harvested, u^*, with $M = 0.9$.

25.6   Beaver Example

We now consider an example about harvesting a beaver population that has caused damage through flooding and destroying trees. This model comes from a paper by Lenhart and Bhat [117]; see that paper for the estimates to justify the needed convergences. In this model, we are only considering the nuisance side of the beavers, not the benefit side. The initial distribution of beavers and some estimates of the parameters came from data from the New York State Department of Environmental Conservation.

The population density of the beaver species is given by the model

$$z_t(x,y,t) - \alpha \Delta z(x,y,t) = z(x,y,t)(a - bz(x,y,t) - u(x,y,t)) \text{ in } \Omega \times (0,T)$$
$$z(x,y,0) = z_0(x,y) \text{ in } \Omega$$
$$z(x,y,t) = 0 \text{ on } \partial\Omega \times (0,T)$$

where α is the constant diffusion coefficient, and a and b are spatially dependent growth parameters. Our control is u, the proportion of population z to be trapped per unit time. The zero boundary conditions imply the unsuitability of the surrounding habitat.

Given the control set

$$U = \{u \in L^2(\Omega \times (0,T)) : 0 \leq u(x,y,t) \leq M\}$$

we seek to minimize the cost functional

$$J(u) = \int_0^T \int_\Omega e^{-rt} \left(\frac{1}{2}\gamma z(x,y,t)^2 + cu(x,y,t)^2 z(x,y,t)\right) dx\, dy\, dt.$$

In this objective functional, $\frac{1}{2}\gamma z^2$ represents the density dependent damage that beavers cause. The $cu^2 z$ term represents the cost of trapping, which is composed of two factors,

$$(cu \text{ unit cost })(uz \text{ amount trapped}).$$

The term e^{-rt} is included for discounted value of the accrued future costs. We find the optimal control p^* that minimizes the objective functional

$$J(u^*) = \min_{u \in U} J(u).$$

The model is very similar to the previous example, so we omit the derivation of the sensitivity equation and just give it below. For a variation l, the resulting PDE for ψ is

$$\psi_t - \alpha \Delta \psi = a\psi - 2bz^*\psi - u^*\psi - lz \text{ on } \Omega \times (0,T)$$
$$\psi = 0 \text{ on } \partial\Omega \times (0,T)$$
$$\psi = 0 \text{ on } \{t = 0\}.$$

The form of the objective functional is slightly different, which affects the adjoint equation. Given an optimal control u^* and corresponding state z^*, the adjoint problem is given by

$$-\lambda_t - \alpha\Delta\lambda = a\lambda - 2bz^*\lambda - r\lambda - u^*\lambda + \gamma z^* + c(u^*)^2 \text{ in } \Omega$$
$$\lambda = 0 \text{ on } \Omega \times \{t = T\}$$
$$\lambda = 0 \text{ on } \partial\Omega \times (0,T).$$

The nonhomogeneous term $\gamma z^* + c(u^*)^2$ in the adjoint equation is from differentiating the integrand $\frac{1}{2}\gamma z^2 + cu^2 z$ of the objective functional (without the discount factor) with respect to the state z. Note again the r term comes from the discount factor. The operator in the adjoint PDE is the adjoint operator of

$$L\psi = \psi_t - \alpha\Delta\psi - a\psi + 2bz^*\psi + u^*\psi.$$

At the optimal control u^*, we again differentiate $J(u)$ with respect to u. The quotient is nonnegative since $J(u^*)$ is the minimum value

$$0 \leq \lim_{\epsilon \to 0^+} \frac{J(u^* + \epsilon l) - J(u^*)}{\epsilon}.$$

We simplify this limit

$$0 \leq \lim_{\epsilon \to 0^+} \int_0^T \int_\Omega e^{-rt}[\frac{1}{2\epsilon}\gamma\left(z^{\epsilon 2} - z^{*2}\right) + c((u^* + \epsilon l)^2 z^\epsilon - u^{*2}z^*))]\, dx\, dt$$
$$= \int_0^T \int_\Omega e^{-rt}[\gamma z^*\psi + cu^{*2}\psi + 2cu^*z^*l]\, dx\, dt$$
$$= \int_0^T \int_\Omega e^{-\delta t}(\psi L^*\lambda + r\psi\lambda + 2cu^*z^*l)\, dx\, dt$$
$$= \int_0^T \int_\Omega e^{-\delta t}(\lambda L\psi + 2cu^*z^*l)\, dx\, dt$$
$$= \int_0^T \int_\Omega e^{-\delta t}(-\lambda l z^* + 2cu^*z^*l)\, dx\, dt$$
$$= \int_0^T \int_\Omega e^{-\delta t}l z^*(-\lambda + 2cu^*)\, dx\, dt.$$

As before, we used

$$\frac{(z^\epsilon)^2 - z^2}{2\epsilon} \to z\psi.$$

Considering the possible variations l, we obtain $u^* = \frac{\lambda}{2c}$ on the interior of the control set (away from the bounds), since z^* is positive inside the domain. Taking the bounds into account, we conclude

$$u^* = \min\left(\max\left(0, \frac{\lambda}{2c}\right), M\right).$$

A forward-backward sweep iteration method with a finite difference scheme was also used to solve this problem. Figures 25.2 and 25.3 show the numerical results of a particular case of the beaver harvesting problem with a ten year time horizon.

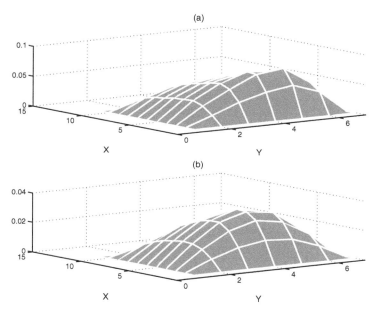

FIGURE 25.2: Proportion to be harvested: (a) at initial time, (b) at $t = 5$ years.

25.7 Predator-Prey Example

Next, we treat an example with a system of PDEs to show how to handle the adjoint system. We consider optimal control of a parabolic system with Neumann boundary conditions. Solutions of the state system represent population densities of the prey and the predator species. The system has Lotka-Volterra type growth terms and local interaction terms between the populations. This example is a somewhat simplified version of the results in [59], where one can find the analysis justification for the results given here. The two controls represent harvesting (actually rates of harvest). The spatial domain is bounded set in $\mathbb{R}^n$.

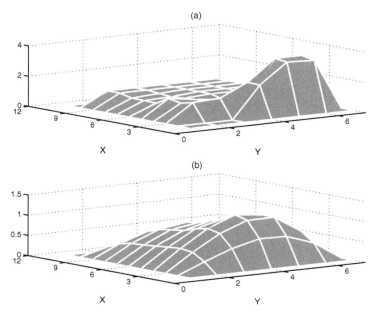

FIGURE 25.3: Population density of beavers: (a) at initial time, (b) at $t = 5$ years.

$$w_t - b_1 \Delta w = (a_1 - d_1 w)w - c_1 wv - u_1 w \quad \text{and}$$
$$v_t - b_2 \Delta v = (a_2 - d_2 v)v + c_2 wv - u_2 v \quad \text{in } Q = \Omega \times (0, T),$$

with initial and boundary conditions

$$w(x, 0) = w_0(x), \quad v(x, 0) = v_0(x) \quad \text{for } x \in \Omega,$$
$$\frac{\partial w}{\partial \nu}(x, t) = 0, \quad \frac{\partial v}{\partial \nu}(x, t) = 0 \quad \text{on } \partial\Omega \times (0, T).$$

Here, the terms and coefficients are

$$
\begin{aligned}
u_1(x,t), u_2(x,t) &= \text{controls} \\
w(x,t) &= \text{prey population } (1^{st} \text{ state variable}) \\
v(x,t) &= \text{predator population } (2^{nd} \text{ state variable}) \\
b_1(x,t), b_2(x,t) &= \text{diffusion coefficients, strictly positive} \\
a_1(x,t), a_2(x,t), d_1(x,t), d_2(x,t) &= \text{standard logistic growth terms} \\
c_1(x,t), c_2(x,t) &= \text{interaction coefficients} \\
\nu(x) &= \text{outward unit normal vector at } x \text{ in } \partial\Omega.
\end{aligned}
$$

Our class of admissible controls is

$$U = \left\{(u_1, u_2) \in L^2(\Omega \times (0,T)) : 0 \leq u_i \leq \Gamma_i \text{ a.e. for } i = 1, 2\right\}.$$

We want to maximize the objective functional $J(u_1, u_2)$ defined by

$$\int_Q \left[K_1 u_1(x,t) w(x,t) + K_2 u_2(x,t) v(x,t) - M_1 u_1^2(x,t) - M_2 u_2^2(x,t)\right] dx\, dt,$$

where $K_1 u_1 w$, $K_2 u_2 v$ represent the revenue of harvesting, and $M_1 u_1^2$, $M_2 u_2^2$ denote the cost of the controls.

Next, we calculate the sensitivities. The mapping

$$(u_1, u_2) \in U \to (w, v)$$

is differentiable, i.e.,

$$\frac{w(u + \epsilon k) - w(u)}{\epsilon} \to \psi_1 \quad \text{and} \quad \frac{v(u + \epsilon k) - v(u)}{\epsilon} \to \psi_2$$

as $\epsilon \to 0$ for any $u = (u_1, u_2) \in U$ and bounded $k = (k_1, k_2)$ such that $(u + \epsilon k) \in U$ for ϵ small. As we did in the previous two examples, one can show that ψ_1, ψ_2 satisfy

$$\begin{aligned}
(\psi_1)_t - b_1 \Delta \psi_1 &= a_1 \psi_1 - 2d_1 \psi_1 w - c_1(w \psi_2 + v \psi_1) - u_1 \psi_1 - k_1 w \quad \text{and} \\
(\psi_2)_t - b_2 \Delta \psi_2 &= a_2 \psi_2 - 2d_2 \psi_2 v + c_2(w \psi_2 + v \psi_1) - u_2 \psi_2 - k_2 v \quad \text{in } Q, \\
\psi_1(x, 0) &= 0 = \psi_2(x, 0) \quad \text{for } x \in \Omega, \\
\frac{\partial \psi_1}{\partial \nu} &= 0 = \frac{\partial \psi_2}{\partial \nu} \quad \text{on } \partial\Omega \times (0, T).
\end{aligned}$$

To derive the optimality system and to characterize the optimal control, we need adjoint variables and adjoints of the operators associated with the ψ_1, ψ_2 system. We write the ψ_1, ψ_2 PDE system as

$$\mathcal{L} \begin{pmatrix} \psi_1 \\ \psi_2 \end{pmatrix} = \begin{pmatrix} -h_1 w \\ -h_2 v \end{pmatrix}$$

where

$$\mathcal{L} \begin{pmatrix} \psi_1 \\ \psi_2 \end{pmatrix} = \begin{pmatrix} \mathcal{L}_1 \psi_1 \\ \mathcal{L}_2 \psi_2 \end{pmatrix} + M \begin{pmatrix} \psi_1 \\ \psi_2 \end{pmatrix}, \quad \begin{pmatrix} \mathcal{L}_1 \psi_1 \\ \mathcal{L}_2 \psi_2 \end{pmatrix} = \begin{pmatrix} (\psi_1)_t - b_1 \Delta \psi_1 \\ (\psi_2)_t - b_2 \Delta \psi_2 \end{pmatrix},$$

and

$$M = \begin{pmatrix} -a_1 + 2wd_1 + c_1v + u_1 & c_1w \\ -c_1v & -a_2 + 2d_2v - c_2w + u_2 \end{pmatrix}.$$

The adjoint PDE system is

$$\mathcal{L}^* \begin{pmatrix} \lambda_1 \\ \lambda_2 \end{pmatrix} = \begin{pmatrix} K_1 u_1 \\ K_2 u_2 \end{pmatrix},$$

where K_1, K_2 are the constants from the objective functional. The adjoint operator is

$$\mathcal{L}^* \begin{pmatrix} \lambda_1 \\ \lambda_2 \end{pmatrix} = \begin{pmatrix} \mathcal{L}_1^* \lambda_1 \\ \mathcal{L}_2^* \lambda_2 \end{pmatrix} + M^\mathsf{T} \begin{pmatrix} \lambda_1 \\ \lambda_2 \end{pmatrix},$$

where M^T denotes the transpose of the matrix. The derivative terms of the adjoint system are the same as in our other two examples. Thus the adjoint PDE equations are

$$\begin{pmatrix} -(\lambda_1)_t - b_1 \Delta \lambda_1 \\ -(\lambda_2)_t - b_2 \Delta \lambda_2 \end{pmatrix} + M^\tau \begin{pmatrix} \lambda_1 \\ \lambda_2 \end{pmatrix} = \begin{pmatrix} K_1 u_1 \\ K_2 u_2 \end{pmatrix}.$$

For the adjoint system, we have the appropriate boundary conditions, namely, zero Neumann conditions and zero final-time conditions. The adjoint system is calculated at the optimal controls $u^* = (u_1^*, u_2^*)$ and corresponding states w^*, v^*. The transversality conditions are

$$\lambda_1(x, T) = 0 \quad \text{and} \quad \lambda_2(x, T) = 0 \quad \text{for } x \in \Omega.$$

We compute the directional derivative of the functional $J(u)$ with respect to u in the direction k at u^*. Since $J(u^*)$ is the minimum value, we have

$$0 \leq \lim_{\epsilon \to 0^+} \frac{J(u^* + \epsilon k) - J(u^*)}{\epsilon}$$

$$= \lim_{\epsilon \to 0^+} \int_W \left[\frac{K_1 u_1(w^\epsilon - w^*) + K_2 u_2(v^\epsilon - v^*) + K_1 k_1 w^\epsilon + K_2 k_2 v^\epsilon}{\epsilon} \right] dx\, dt$$

$$- \frac{1}{\epsilon} \int_Q \left[M_1((u_1^* + \epsilon k_1)^2 - (u_1^*)^2) + M_2((u_2^* + \epsilon k_2)^2 - (u_2^*)^2) \right] dx dt$$

$$= \int_Q (\psi_1, \psi_2) \begin{pmatrix} K_1 u_1 \\ K_2 u_2 \end{pmatrix} dx\, dt$$

$$+ \int_Q k_1(K_1 w^* - 2M_1 u_1^*) + k_2(K_2 v^* - 2M_2 u_2^*)\, dx\, dt$$

$$= \int_Q \left[b_1 \nabla \lambda_1 \cdot \nabla \psi_1 + b_2 \nabla \lambda_2 \cdot \nabla \psi_2 + (\psi_1, \psi_2) M^\tau \begin{pmatrix} \lambda_1 \\ \lambda_2 \end{pmatrix} \right] dx\, dt$$

$$- \int_Q (\lambda_1)_t \psi_1 + (\lambda_2)_t \psi_2 \, dx\, dt$$

$$+ \int_Q k_1(K_1 w^* - 2M_1 u_1^*) + k_2(K_2 v^* - 2M_2 u_2^*)\, dx dt$$

$$= \int_Q (\lambda_1, \lambda_2) \begin{pmatrix} -k_1 w^* \\ -k_2 v^* \end{pmatrix} dx\, dt$$

$$+ \int_Q k_1(K_1 w^* - 2M_1 u_1^*) + k_2(K_2 v^* - 2M_2 u_2^*)\, dx dt$$

$$= \int_Q (-k_1 \lambda_1 w^* - k_2 \lambda_2 v^*)\, dx\, dt$$

$$+ \int_Q [k_1(K_1 w^* - 2M_1 u_1^*) + k_2(K_2 v^* - 2M_2 u_2^*)]\, dx dt$$

$$= \int_Q k_1 \left(K_1 w^* - \lambda_1 w^* - 2M_1 u^* \right) + k_2 \left(K_2 v^* - \lambda_2 v^* - 2M_2 u_2^* \right) dx\, dt.$$

From the above calculation, we obtain the characterization of the optimal control pair:

$$u_1^*(x) = \min\left(\Gamma_1, \max\left(\frac{1}{2M_1}(K_1 - \lambda_1) w^*, 0 \right) \right),$$

$$u_2^*(x) = \min\left(\Gamma_2, \max\left(\frac{1}{2M_2}(K_2 - \lambda_2) v^*, 0 \right) \right).$$

25.8 Identification Example

This example illustrates that in some cases, optimal control theory can be applied to an identification problem. We want to identify the "unknown" coefficient of the interaction term in a predator-prey system with a Neumann boundary condition in a two-dimensional bounded spatial domain. Let w represent the population concentration of prey and v the predator concentration. The system has local interaction terms representing a predator-prey situation, and the prey equation has Lotka-Voltera type growth term. Our aim is to identify the interaction coefficient by optimal control techniques. We refer the reader to the book by Banks and Kunisch [8] for more background and other techniques for estimation problems.

The state system is

$$w_t(x,t) - b_1 \Delta w(x,t) = \big(a_1(x,t) - d(x,t)w(x,t)\big)w(x,t) - u(x)w(x,t)v(x,t),$$
$$v_t(x,t) - b_2 \Delta v(x,t) = -a_2(x,t)v(x,t) + u(x)w(x,t)v(x,t),$$

in $Q = \Omega \times (0,T)$, with initial and boundary conditions

$$w(x,0) = w_0(x), \quad v(x,0) = v_0(x) \quad \text{for } x \in \Omega,$$
$$\frac{\partial w}{\partial \nu}(x,t) = 0, \quad \frac{\partial v}{\partial \nu}(x,t) = 0 \quad \text{on } \partial \Omega \times (0,T).$$

Here, the terms and coefficients are

$$u(x) = \text{coefficient of the interaction term to be identified}$$
$$w(x,t) = \text{prey population (first state variable)}$$
$$v(x,t) = \text{predator population (second state variable)}$$
$$b_1, b_2 = \text{diffusion coefficients}$$
$$a_1(x,t), a_2(x,t), d(x,t) = \text{standard logistic growth terms}$$
$$\nu(x) = \text{outward unit normal vector at } x \text{ in } \partial\Omega.$$

Our class of admissible interaction coefficients (controls) is

$$U = \{u \in L^2(\Omega) : 0 \leq u \leq M \text{ a.e. in } \Omega\}.$$

We want to minimize the objective functional

$$J(u) = \int_W \frac{1}{2}\Big[\big(w(x,t) - z_1(x,t)\big)^2 + \big(v(x,t) - z_2(x,t)\big)^2\Big]\,dx\,dt + \frac{\beta}{2}\int_\Omega u(x)^2\,dx,$$

where $W \subset Q$, the area of W is positive, and z_1, z_2 are observations of the prey and predator populations. Namely, given partial (and perhaps noisy) observations z_1, z_2 of the true solution w, v in a subdomain W of Q, we seek to "identify" the $u(x)$ which best matches the model to the data. Note the control u is only a function of x, while our PDE system has space and time variables.

The second term of the objective functional is artificial; we have added it to prevent the problem from being linear in the control. In practice, we choose β very small, so the emphasis of the problem lies in minimizing the closeness of the states to the observed data. This method of solving this identification problem is based on Tikhonov's regularization. Our eventual estimate for u will depend on the choice of β.

The mapping $u \in U \to (w, v)$ is differentiable, i.e.,

$$\frac{w(u+\epsilon k) - w(u)}{\epsilon} \to \psi_1 \quad \text{and} \quad \frac{v(u+\epsilon k) - v(u)}{\epsilon} \to \psi_2$$

as $\epsilon \to 0$ for any $u \in U$ and bounded k such that $(u + \epsilon k) \in U$ for ϵ small. As we did in the previous two examples, one can show that ψ_1, ψ_2 satisfy

$$(\psi_1)_t - b_1 \Delta \psi_1 = a_1 \psi_1 - 2d\psi_1 w - u(w\psi_2 + v\psi_1) - kwv \quad \text{and}$$
$$(\psi_2)_t - b_2 \Delta \psi_2 = -a_2 \psi_2 + u(w\psi_2 + v\psi_1) + kwv \quad \text{in } Q,$$
$$\psi_1(x, 0) = 0 = \psi_2(x, 0) \quad \text{for } x \in \Omega,$$
$$\frac{\partial \psi_1}{\partial \nu} = 0 = \frac{\partial \psi_2}{\partial \nu} \quad \text{on } \partial\Omega \times (0, T).$$

To derive the optimality system and to characterize the optimal control, we need adjoint variables and adjoint operators associated with the ψ_1, ψ_2 system. We write the ψ_1, ψ_2 PDE system as

$$\mathcal{L} \begin{pmatrix} \psi_1 \\ \psi_2 \end{pmatrix} = \begin{pmatrix} -kwv \\ kwv \end{pmatrix}$$

where

$$\mathcal{L} \begin{pmatrix} \psi_1 \\ \psi_2 \end{pmatrix} = \begin{pmatrix} \mathcal{L}_1 \psi_1 \\ \mathcal{L}_2 \psi_2 \end{pmatrix} + M \begin{pmatrix} \psi_1 \\ \psi_2 \end{pmatrix},$$

$$\begin{pmatrix} \mathcal{L}_1 \psi_1 \\ \mathcal{L}_2 \psi_2 \end{pmatrix} = \begin{pmatrix} (\psi_1)_t - b_1 \Delta \psi_1 \\ (\psi_2)_t - b_2 \Delta \psi_2 \end{pmatrix}, \quad \text{and } M = \begin{pmatrix} -a_1 + 2wd + uv & uw \\ -uv & a_2 - uw \end{pmatrix}.$$

We define the adjoint PDE system as

$$\mathcal{L}^* \begin{pmatrix} \lambda_1 \\ \lambda_2 \end{pmatrix} = \begin{pmatrix} w - z_1 \\ v - z_2 \end{pmatrix} \chi_W,$$

where χ_W is the characteristic function of the set W,

$$\mathcal{L}^* \begin{pmatrix} \lambda_1 \\ \lambda_2 \end{pmatrix} = \begin{pmatrix} \mathcal{L}_1^* \lambda_1 \\ \mathcal{L}_2^* \lambda_2 \end{pmatrix} + M^{\mathrm{T}} \begin{pmatrix} \lambda_1 \\ \lambda_2 \end{pmatrix}$$

and

$$\begin{pmatrix} \mathcal{L}_1^* \lambda_1 \\ \mathcal{L}_2^* \lambda_2 \end{pmatrix} = \begin{pmatrix} -(\lambda_1)_t - b_1 \Delta \lambda_1 \\ -(\lambda_2)_t - b_2 \Delta \lambda_2 \end{pmatrix}.$$

The components of the RHS of this system are the derivatives of the first integrand in with objective functional with respect to each state. We rewrite the objective functional as

$$\frac{1}{2} \int_Q \left[(w(u) - z_1)^2 + (v(u) - z_2)^2 \right] \chi_W \, dx \, dt + \frac{\beta}{2} \int_\Omega u(x)^2 \, dx.$$

For the adjoint system, we have the appropriate boundary conditions, zero Neumann conditions and zero final-time conditions. For a fixed β, the adjoint system is calculated at the optimal control u^* and corresponding states w^*, v^*

$$\mathcal{L}_1^* \lambda_1 = (w^* - z_1)\chi_W + a_1 \lambda_1 - 2dw^* \lambda_1 - u^*(v^* \lambda_1 - v^* \lambda_2) \quad \text{in } Q,$$
$$\frac{\partial \lambda_1}{\partial \nu} = 0 \quad \text{on } \partial\Omega \times (0,T),$$

and

$$\mathcal{L}_1^* \lambda_2 = (v^* - z_2)\chi_W - a_2 \lambda_2 + u^*(w^* \lambda_2 - w^* \lambda_1) \quad \text{in } Q,$$
$$\frac{\partial \lambda_2}{\partial \nu} = 0 \quad \text{on } \partial\Omega \times (0,T).$$

The transversality conditions are

$$\lambda_1(x,T) = 0 \quad \text{and} \quad \lambda_2(x,T) = 0 \quad \text{for } x \in \Omega.$$

We compute the directional derivative of the functional $J(u)$ with respect to u in the direction k at u^*. Since $J(u^*)$ is the minimum value, we have

$$0 \leq \lim_{\epsilon \to 0^+} \frac{J(u^* + \epsilon k) - J(u^*)}{\epsilon}$$

$$= \frac{1}{2} \lim_{\epsilon \to 0^+} \int_W \left[\frac{(w^\epsilon - z_1)^2 - (w^* - z_1)^2}{\epsilon} + \frac{(v^\epsilon - z_2)^2 - (v^* - z_2)^2}{\epsilon} \right] dx\, dt$$

$$+ \frac{\beta}{2\epsilon} \int_\Omega \left[(u^* + \epsilon k)^2 - (u^*)^2 \right] dx$$

$$= \int_Q (\psi_1, \psi_2) \begin{pmatrix} (w^* - z_1)\chi_W \\ (v^* - z_2)\chi_W \end{pmatrix} dx\, dt + \beta \int_\Omega u^* k\, dx$$

$$= \int_Q \left[b_1 \nabla \lambda_1 \cdot \nabla \psi_1 + b_2 \nabla \lambda_2 \cdot \nabla \psi_2 + (\psi_1, \psi_2) M^T \begin{pmatrix} \lambda_1 \\ \lambda_2 \end{pmatrix} \right] dx\, dt$$

$$+ \int_Q -[(\lambda_1)_t \psi_1 + (\lambda_2)_t \psi_2]\, dx dt + \beta \int_\Omega u^* k\, dx$$

$$= \int_Q (\lambda_1, \lambda_2) \begin{pmatrix} -k w^* v^* \\ k w^* v^* \end{pmatrix} dx\, dt + \beta \int_\Omega u^* k\, dx$$

$$= \int_Q (-k\lambda_1 w^* v^* + k\lambda_2 w^* v^*)\, dx\, dt + \beta \int_\Omega u^* k\, dx$$

$$= \int_\Omega k(x) \left(\beta u^* + \int_0^T (-\lambda_1 w^* v^* + k \lambda_2 w^* v^*)\, dt \right) dx.$$

From the above calculation, we obtain the characterization of the optimal control (explicitly showing the β dependence):

$$u_\beta(x) = \min\left(M, \max\left(\frac{1}{\beta} \int_0^T (\lambda_1 - \lambda_2) w_\beta v_\beta\, dt, 0 \right) \right). \tag{25.1}$$

We illustrate a numerical example in which the exact $u(x) = x + 1/4$ and use the observation set $W = (0, 1/2) \times (0, 1/2)$. The observation functions were

$$z_1(x, t) = e^{-t}(2 + \cos \pi x), \quad z_2(x, t) = e^{-t}(2 - \cos \pi x).$$

In Figure 25.4, we can see the actual value u versus the approximate optimal control u_β^* in the subdomain $(0, 1/2)$ and with $\beta = 0.05$.

25.9 Controlling Boundary Terms

In control of PDEs, one could choose boundary source terms or coefficients as controls. Consider a parabolic PDE system with Neumann boundary conditions

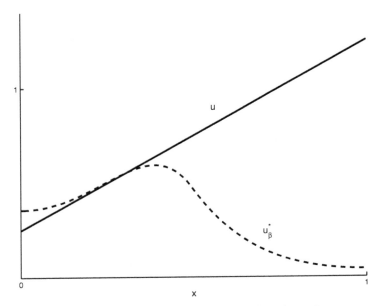

FIGURE 25.4: Exact u and computed approximation u_β^*.

$$w_t - \Delta w = f \text{ on } Q = \Omega \times (0, T)$$
$$w = w_0(x) \text{ on } \Omega \times \{0\}$$
$$\frac{\partial w}{\partial \nu} = u \text{ on } \Omega \times (0, T).$$

We take our control to be the flux on the boundary. The source term f is a given bounded function. Our control set is

$$U = \left\{ u \in L^2(\partial\Omega \times (0,T)) : 0 \le u(x,t) \le M \right\}.$$

We seek to adjust the flux control to drive the solution toward a desired profile and minimize the cost of the control. The objective functional is

$$J(u) = \int_Q \frac{1}{2}(w(x,t) - z(x,t))^2 \, dx \, dt + \frac{B}{2} \int_{\partial\Omega \times (0,T)} u(x,t)^2 \, dsdt,$$

where z is a given bounded function. This example is a simple version of the work in [46]. Since we have chosen a linear PDE, the sensitivity equation would have the same structure as the state PDE

$$\psi_t - \Delta\psi = 0 \text{ on } Q = \Omega \times (0,T)$$
$$\psi = 0 \text{ on } \Omega \times \{0\}$$
$$\frac{\partial \psi}{\partial \nu} = k \text{ on } \Omega \times (0,T),$$

where ψ is the derivational derivative of w with respect to u in the direction l. The adjoint equation becomes

$$-\lambda_t - \Delta\lambda = w - z \text{ on } Q = \Omega \times (0,T)$$
$$\lambda = 0 \text{ on } \Omega \times \{T\}$$
$$\frac{\partial \lambda}{\partial \nu} = 0 \text{ on } \Omega \times (0,T).$$

When the objective functional is differentiated, we obtain

$$0 \leq \lim_{\epsilon \to 0^+} \frac{J(u^* + \epsilon k) - J(u^*)}{\epsilon}$$
$$= \frac{1}{2} \lim_{\epsilon \to 0^+} \int_Q \left[\frac{(w^\epsilon - z)^2 - (w^* - z)^2}{\epsilon} \right] dx\, dt$$
$$+ \frac{B}{2\epsilon} \int_{\partial\Omega \times (0,T)} \left[(u^* + \epsilon k)^2 - (u^*)^2 \right] ds dt$$
$$= \int_Q \psi(w^* - z)\, dx\, dt + B \int_{\partial\Omega \times (0,T)} u^* k\, ds\, dt$$
$$= \int_Q [-\psi\lambda_t + \nabla\lambda\nabla\psi]\, dx\, dt + B \int_{\partial\Omega \times (0,T)} u^* k\, ds\, dt$$
$$= \int_{\partial\Omega \times (0,T)} k\, (\lambda + Bu^*)\, ds\, dt.$$

Thus the optimal control characterization is

$$u^*(x) = \min\left(M, \max\left(\frac{-\lambda}{B}, 0 \right) \right).$$

Note that the control appears linearly in this PDE problem and quadratically in the objective functional. Thus, it is a PDE generalization of the linear quadratic regulator problem treated in chapter 12. See the book by Lasiecka and Triggiani for results for such PDE control problems [111, 112]. For an example of a system with control on the boundary terms, see [119].

25.10 Exercises

Exercise 25.1 Consider the following predator-prey problem, which has a slightly different model than in Section 25.7. Find the PDE system and boundary conditions for the sensitivities and the adjoint, and calculate the characterization of the optimal control.

$$w_t - b_1 \Delta w = (a_1 - d_1 w)w - c_1 wv - u_1 w \quad \text{and}$$
$$v_t - b_2 \Delta v = -a_2 v + c_2 wv - u_2 v \quad \text{in } Q = \Omega \times (0, T),$$

with initial and boundary conditions

$$w(x,0) = w_0(x), \quad v(x,0) = v_0(x) \quad \text{for } x \in \Omega,$$
$$\frac{\partial w}{\partial \nu}(x,t) = 0, \quad \frac{\partial v}{\partial \nu}(x,t) = 0 \quad \text{on } \partial\Omega \times (0, T).$$

Our class of admissible controls is

$$U = \left\{ (u_1, u_2) \in L^2(\Omega \times (0, T)) : 0 \le u_i \le \Gamma_i \text{ a.e. for } i = 1, 2 \right\}.$$

Maximize the objective functional $J(u_1, u_2)$ defined by

$$\int_Q \left[K_1 u_1(x,t) w(x,t) + K_2 u_2(x,t) v(x,t) - M_1 u_1^2(x,t) - M_2 u_2^2(x,t) \right] dx\, dt.$$

Exercise 25.2 Find the equations for the sensitivity and the adjoint and the characterization of the optimal control. You can check your results in [123].

$$w_t - \Delta w = f \text{ on } Q = \Omega \times (0, T)$$
$$w = w_0(x) \text{ on } \Omega \times \{0\}$$
$$\frac{\partial w}{\partial \nu} + uw = 0 \text{ on } \partial\Omega \times (0, T).$$

Our control set is

$$U = \left\{ u \in L^2(\partial\Omega \times (0, T)) : 0 \le u(x,t) \le M \right\}.$$

The objective functional is

$$J(u) = \int_Q \frac{1}{2}(w(x,t) - z(x,t))^2 \, dx \, dt + \frac{B}{2} \int_{\partial\Omega \times (0,T)} u(x,t)^2 \, ds \, dt.$$

Chapter 26

Other Approaches and Extensions

In this final chapter, we present briefly some other topics in optimal control and optimization problems of biological models. Alternative approaches to solutions are discussed and some types of underlying dynamics which have not been treated here are mentioned.

Dynamic Programming

So far, we have solved optimal control problems via the Maximum Principle. However, this is not the only method for such problems. Another approach, called dynamic programming, was developed by Richard Bellman. Dynamic programming provides a procedure for determining the optimal combination of decisions. It uses the principle of optimality to verify that the value function satisfies a partial differential equation with appropriate boundary conditions. We first present this approach as applied to optimal control of differential equations, and then we present some background in the application of this approach to discrete models.

Recall from Section 2.3, the principle of optimality roughly says: an optimal control for a given problem must also be optimal for the reduced problem, where the initial conditions are shifted forward along the optimal path. In our basic optimal control problem for ordinary differential equations, we use $u(t)$ for the control and $x(t)$ for the state. The problem is cast as

$$\max_u \int_{t_0}^{t_1} f(t, x(t), u(t))\, dt$$

$$\text{subject to} \quad x'(t) = g(t, x(t), u(t)), \quad x(t_0) = x_0.$$

Also, recall the value function V

$$V(x_0, t_0) := \max_u \int_{t_0}^{t_1} f(t, x(t), u(t))\, dt$$

$$\text{subject to} \quad x'(t) = g(t, x(t), u(t)), \quad x(t_0) = x_0.$$

Note that $V(x(t_1), t_1) = 0$. Fix $\Delta t > 0$ small, and break this integral into two pieces,

$$V(x_0, t_0) = \max_u \left[\int_{t_0}^{t_0+\Delta t} f(t, x(t), u(t))\, dt + \int_{t_0+\Delta t}^{t_1} f(t, x(t), u(t))\, dt \right].$$

Using the principle of optimality, the optimal control for the full problem should be optimal for the problem starting at $t_0 + \Delta t$. So,

$$V(x_0, t_0) = \max_u \left[\int_{t_0}^{t_0+\Delta t} f(t, x(t), u(t))\, dt + V(x(t_0 + \Delta t), t_0 + \Delta t) \right]$$

where the maximum is taken over controls u on the interval $[t_0, t_0 + \Delta t]$. We can assume Δt is so small the integrand is approximately constant. Further, suppose we can expand the second term on the right hand side into a Taylor series (assuming enough regularity on V). Then,

$$V(x_0, t_0) = \max_u [f(t_0, x_0, u)\Delta t + V(x_0, t_0) + V_t(x_0, t_0)\Delta t + V_x(x_0, t_0)\Delta x + h.o.t.],$$

where $h.o.t.$ represents the higher order terms from the Taylor series, and $\Delta x = x(t_0 + \Delta t) - x(t_0)$. Dividing by Δt and letting $\Delta t \to 0$ the dynamic programming PDE or the Hamilton-Jacobi-Bellman equation can be obtained. Namely,

$$-V_t(x, t) = \max_u [f(t, x, u) + V_x(x, t)g(t, x, u)]$$

with boundary condition $V(x(t_1), t_1) = 0$. Note this PDE holds for any state position x and t in the time interval. It has dynamics flowing backward from the final time t_1. See the theory of viscosity solutions developed for solutions to such equations [39], especially when V is not differentiable.

If, in your model, the decision times can be divided into discrete times and the range space of the states has a finite number of values, then dynamic programming for discrete models is a reasonable tool. This approach also works well for discrete time problems (or discrete in other underlying variables) since the optimal policy at stage n (or time step n) can be obtained from the optimal policy at stage $n+1$. This iterative backward solving is the crucial tool. Frequently the range space of the control variable is a finite set. We point out the interesting work of Mangel and Clark [34, 136] and Houston and McNamara [86] in this area of dynamic programming. Mangel and Clark call their models "dynamic state variable models," which are individual optimization models of behavior and are well suited to empirical studies based on field or laboratory data [34, 136]. Some of their problems include foraging and fattening strategies for willow tits, overwintering foraging by juvenile salmon, and predicting flowering in thistles. The organism and the environment can be linked through the expected lifetime success of the organism. Important issues are whether behaving optimally is important for the organism and evaluating the fitness of alternative strategies. This optimization approach had a big impact on the field of behavioral ecology, and the ease of coding the algorithm and the applicability to appropriate models were crucial. Using a continuous approach as opposed to this discrete approach depends on the situation to be modeled, the number of decisions to be made, and the possible range values of the state variables.

We also call attention to the idea of stochastic dynamic programming [161]. Possingham and his collaborators have worked extensively on ecological applications of this approach; we cite only one of many such papers [160].

Linear and Nonlinear Programming

A third approach to optimal control problems is linear and nonlinear programming. This involves two steps. First, using linear and nonlinear programming to formulate the control problem or starting with a standard optimal control formulation using ODEs or discrete equations. Second, using nonlinear programming to solve the problem.

First, we point out novel examples of formulating spatial optimization problems using linear and nonlinear programming and then solving them with corresponding programming techniques. The books by Hof and Bevers [84, 85] present ideas and methods for directly optimizing the spatial layout of management actions across an ecosystem landscape. The spatial relationships can be simple static or involve spatial autocorrelation and dynamic changes through time. In [85], the two ways of depicting spatial options are cellular grids and geometric shapes. In the cellular grid problems, the control variables are defined for each cell and irregular shapes are approximated by aggregation of cells. In the geometric shapes case, the control variables define the size and location of the geometric shapes. In many of their examples, the underlying ecological relationships are represented in discrete difference equations with constraints on the control and state variables. The methods for the numerical solutions include the simplex algorithm, "integer-friendly" linear mixed-integer programs, and generalized reduced gradient algorithms. The applications are varied and range from ferret reintroduction in South Dakota to strategies for controlling wildfires. Note that these examples are discrete in time and space. Choosing such an approach depends on the scale of your model.

Second, there are a variety of methods to solve optimal control problems via programming, many of which involve discretizing the ODE state equations and the objective functional and using an optimization approach. The book by Betts [17] concentrates on numerical methods for solving optimal control problems, using a sequence of simpler finite dimensional subproblems. His examples are engineering oriented, but the description of their approach, "first discretize and then optimize," is worthwhile. The book by Gregory and Lin [71] is also a useful reference for numerical techniques involving difference equations and then generalized spline matrices.

Control versus Optimal Control

We have not treated the area of "control," which deals with the following issues:

- Controllability
  (use controls to steer system from one position to another)

- Observability
  (deduce system information from control input and observation output)

- Stabilization
  (implement controls to force stability)

These are frequently considered from the viewpoint of feedback control [30, 87, 116], meaning the optimal control was a function of the state (and not the adjoint), as in section 12.2. Stabilization results can be obtained by using Laplace transforms of linear state systems and calculating the poles of the transfer function, which takes the transformed input function to the output function [148]. We also did not consider infinite time horizon problems, which are important in economics applications [33, 87, 100, 124, 167].

The area of geometric control theory is also interesting. It deals with the idea of reachable sets and trajectories of the state systems, including the case of having a terminal surface (or manifold) for endpoints of the trajectories. The application of Lie algebras and brackets to the controllability of certain ODE systems has applications, for example, in robotics [19, 99, 151, 164, 172]. Inherently, Lie brackets can measure the degree of noncommutativity of operators, and this property can enable a system to reach more positions in the state space.

Alternate Dynamic Features

We have treated systems of ordinary differential equations and partial differential equations with continuous dynamics, in addition to discrete equations. There are many useful other possibilities for features in the dynamics of the state variables. One can easily imagine differential equations with delays [41, 76, 143, 150], like a drug or vaccine treatment that has a delay before its effect starts. See [4, 21, 100] for the necessary conditions of optimal control and [77] for feedback control for delay differential equations.

It is also possible to have dynamics with jumps in the state variables. For example, consider a model of the seasonal dynamics of ticks and their hosts. The hosts could have continuous dynamics, but the ticks arrest their development in the winter. Thus, the ticks have dynamics in the summers and the initial condition for one summer could depend on the previous summer, leading to possible discontinuities or jumps in the dynamics [69]. In many engineering applications, jumps in the state dynamics and in the controls have lead to new results [63, 144]. Systems with a mixture of different dynamic features like discrete, continuous, and jumps are often called hybrid systems [75, 121].

Integrodifferential equations have been used successfully to model populations with distinct growth and dispersal stages. A simple such model would be

$$N_{t+1}(x) = \int_\Omega k(x,y) f(N_t(y)) \, dy,$$

where f is the growth function, and $k(x,y)$ is the dispersal kernel. For a variety of plant and animal population models with integrodifferential equations, see the work of Kot, Lewis, and Neubert [108, 127, 153]. These systems are a type of hybrid system, and recently, the optimal control techniques have been developed, combining ideas from the discrete version of Pontryagin's Maximum Principle and techniques from optimal control of PDEs [91, 94]. For an application to a plant pathogen model in crops, see [67].

The underlying dynamics of an optimal control problem could be stochastic differential equations (SDEs) due to randomness in the habitat or demographic stochasticity features. There are different types of "calculus" for such SDEs. The most common type is Ito calculus; for more information see [62, 65, 88]. As in the optimal control of ODEs, where the adjoint dynamics have final time data, the stochastic differential equations (SDEs) case can be solved with backwards SDEs [134]. Another interesting type of system involving randomness is piecewise-deterministic processes with deterministic ODE dynamics between random jumps [43].

In the discrete time case, further extensions can be considered beyond what was treated here. For example, the state equation could depend on the control values at the two previous time steps, not just the immediately preceding steps. See [96, 97] for an application in improving the chest compression-decompression pattern in the standard technique of cardiopulmonary resuscitation (CPR) using a seven compartment circulation model. Models that are discrete in space and in time are amenable to the techniques treated here.

Sometimes state constraints can be incorporated in the dynamics of the system, as in having a lower or upper obstacle on the state values. One could restrict harvesting so that the state population stays above an enforced lower bound. In this case, the state ODE or PDE equation would be an inequality. When the state is strictly above its obstacle, equality would hold. This framework is called variational inequalities [103], and the optimal control framework was developed by Barbu [10, 11]. In general, state constraints are much more difficult to handle than control constraints, whether or not the state constraints are incorporated into the dynamics [50]. See the survey paper by Hartl, Sethi, and Vickson [80].

Optimizing Parameters

The idea of optimal control is easily extended to problems where parameters are to be optimized. Consider this simple illustrative example:

$$\max_{u \in \mathbb{R}} \left[u^2 + \int_0^T f(t, x(t)) \, dt \right]$$

$$\text{subject to} \quad x'(t) = g(t, x(t), u), \ x(0) = x_0,$$

where the parameter u to be controlled (or optimized over) is taken from the real numbers. If we differentiate the objective functional $J(u)$ with respect to u at u^* and choose the adjoint function to satisfy

$$\lambda'(t) = -\Big[f_x(t, x(t)) + \lambda(t)g_x(t, x(t), u)\Big], \ \lambda(T) = 0,$$

then one obtains

$$0 = 2u^* + \int_0^T \lambda(t) g_u(t, x(t), u) \, dt.$$

We conclude that

$$u^* = -\frac{1}{2} \int_0^T \lambda(t) g_u(t, x(t), u) \, dt.$$

If the only goal is to estimate or optimize a parameter, then other techniques may be preferable. However, for problems where a parameter and a control function are to be optimized simultaneously, then the above method should be considered [101].

There is another technique for estimating parameters, frequently called *data assimilation*. It uses the adjoint equations to approximate the gradient of the objective functional, which would measure the distance from the state output from the observations [15, 42, 72]. That gradient information can be used in an optimization method such as steepest descent [102]. Consider a problem viewing u as a vector of parameters to be estimated:

$$\max_{u \in \mathbb{R}^n} \left[\int_0^T f(t, x(t), u) \, dt \right]$$

$$\text{subject to} \quad x'(t) = g(t, x(t), u), \ x(0) = x_0.$$

Define the adjoint function by

$$\lambda'(t) = -\Big[f_x(t, x(t), u) + \lambda(t)g_x(t, x(t), u)\Big], \ \lambda(T) = 0.$$

Differentiating the objective functional gives

$$\lim_{\epsilon \to 0} \frac{J(u + \epsilon h) - J(u)}{\epsilon} = \int_0^T [f_u(t, x(t), u) + \lambda(t)g_u(t, x(t), u)]h(t) \, dt,$$

and we obtain the gradient of J,

$$\nabla J(u)(t) = f_u(t, x(t), u) + \lambda(t)g_u(t, x(t), u).$$

The state solution at a particular parameter vector u and the corresponding adjoint solution can be used to calculate this gradient. The adjoint equation is used, but not necessarily evaluated at the optimal control.

Final Remarks

This book has mainly focused on the positive side of optimal control theory; that is, methods of solving such problems and usefulness in applications. However, it should be pointed out that many things can go wrong, and our methods sometimes fail. We call your attention to a book with counterexamples in optimal control [168] and to a section of Macki and Strauss [135] on "three discouraging examples" about existence of optimal control results. We suggest the volume by Blondel and Megretski [18] on open problems in control and systems theory.

Throughout, we have focused on the case of piecewise continuous controls. Of interest is the theory of *chattering controls*, those controls which bounce between the upper and lower bounds an infinite number of times in a finite time interval; see the book by Zelikin and Borisov [183].

More control work and biological applications in (partial) functional differential equations [181] would be worthwhile. Also, there are many types of biological models that have not been investigated from an optimal control or optimization viewpoint. In particular, we mention cellular automata and individual-based (or agent-based) models [73]. Further work on multi-drug treatments in disease and cancer models is needed, including linking epidemiology and immunology models. Multilevel models (sometimes called multi-models) of diseases or population interactions, which link the individual, local, and regional levels would be interesting to explore. Also, some biocontrol problems for invasive populations has been investigated via scenario analysis [81]. We hope that optimal control techniques can be seen as a viable alternative.

References

[1] E. Ackerman, L. Gatewood, J. Rosevar, and G. Molnar. *Concepts and Models of Biomathematics*. Marcel Dekker, New York, 1969.

[2] N. U. Ahmed and K. L. Teo. *Optimal Control of Distributed Parameter Systems*. North Holland, New York, 1981.

[3] G. M. Aly. The computation of optimal singular control. *International Journal of Control*, 28(5):681–8, 1978.

[4] T. S. Angell. Existence theorems for optimal control problems involving functional differential equations. *Journal of Optimization Theory and Applications*, 7:149–69, 1971.

[5] Sebastian Anita. *Analysis and Control of Age-Dependent Population Dynamics*. Kluwer, Dordrecht, Netherlands, 2000.

[6] U. Ascher, R. Mattheij, and R. Russell. *Numerical Solutions of Boundary Value Problems for Ordinary Differential Equations*. SIAM, Philadelphia, 1995.

[7] U. Ascher and L. R. Petzold. *Computer Methods for Ordinary Differential Equations and Differential-Algebraic Equations*. SIAM, Philadelphia, 1998.

[8] H. T. Banks and K. Kunisch. *Estimation Techniques for Distributed Parameters Systems*. Birkhäuser, Boston, 1999.

[9] H. T. Banks, H-D. Kwon, J. Toivanen, and H. T. Tran. A state-dependent Riccati equation-based estimator approach for HIV feedback control. *Optimal Control Applications and Methods*, 27, 2006.

[10] V. Barbu. *Analysis and Control of Nonlinear Infinite Dimensional Systems*. Academic Press, New York, 1993.

[11] V. Barbu. *Mathematical Methods in Optimization of Differential Systems*. Kluwer Academic Publishers, Dordrecht, 1994.

[12] Martino Bardi and Italo Capuzzo-Dolcetta. *Optimal Control and Viscosity Solutions of Hamilton-Jacobi-Bellman Equations*. Birkhäuser, Boston, 1997.

[13] H. Behncke. Optimal control of deterministic epidemics. *Optimal Control Applications and Methods*, 21:269–85, 2000.

[14] David J. Bell and David H. Jacobson. *Singular Optimal Control Problems*. Academic Press, New York, 1975.

[15] A. F. Bennett. *Inverse Modeling of the Ocean and Atmosphere*. Cambridge University Press, Cambridge, 2002.

[16] L. D. Berkovitz. *Optimal Control Theory*. Springer-Verlag, New York, 1974.

[17] John. T. Betts. *Practical Methods for Optimal Control Using Nonlinear Programming*. SIAM, Philadelphia, 2001.

[18] Vincent D. Blondel and Alexandre Megretski. *Unsolved Problems in Mathematical Systems and Control Theory*. Princeton University Press, Princeton, 2004.

[19] R. W. Brockett. *Finite Dimensional Linear Systems*. Wiley, New York, 1970.

[20] Arthur E. Bryson, Jr. and Yu-chi Ho. *Applied Optimal Control*. Ginn and Company, Waltham, Massachusetts, 1969.

[21] J. J. Budelis and A. E. Bryson, Jr. Some optimal control results for differential-difference systems. *IEEE Transactions on Automatic Control*, 15:237–41, 1970.

[22] Richard L. Burden and J. Douglas Faires. *Numerical Analysis*. PWS-KENT Publishing Company, Boston, 2004.

[23] John Butcher. On the convergence of numerical solutions of ordinary differential equations. *Math. Comp.*, 20:1–10, 1966.

[24] John Butcher. A multistep generalization of Runge-Kutta methods with four or five stages. *Journal of the ACM*, 14:84–99, 1967.

[25] S. Butler, D. Kirschner, and S. Lenhart. Optimal control of chemotherapy affecting the infectivity of HIV. *Advances in Mathematical Population Dynamics - Molecules, Cells and Man*, 6:557–69, 1997.

[26] J. Canada, J. L. Gamez, and J. A. Montero. A study of an optimal control problem for diffusive nonlinear elliptic equations of logistic type. *SIAM J. of Control Optim.*, 36:1171–89, 1998.

[27] R. S. Cantrell and C. Cosner. *Spatial Ecology via Reaction-Diffusion Equations*. Wiley, West Sussex, England, 2003.

[28] Hal Caswell. *Matrix Population Models: Construction, Analysis and Interpretation*. Sinauer Press, Sunderland, MA., 2001.

[29] Lamberto Cesari. *Optimization Theory and Applications: Problems with Ordinary Differential Equations*. Springer Verlag, New York, 1983.

[30] Chi-Tsong Chen. *Linear System Theory and Design.* Oxford University Press, New York, 1999.

[31] Yaobin Chen and Jian Huang. A numerical algorithm for singular optimal control synthesis using continuation methods. *Optimal Control Applications and Methods,* 15:223–36, 1994.

[32] Ward Cheney and David Kincaid. *Numerical Mathematics and Computing.* Thomson, Belmont, California, 2004.

[33] C. W. Clark. *Mathematical Bioeconomics: The Optimal Management of Renewable Resources.* Wiley, New York, 1990.

[34] C. W. Clark and Marc Mangel. *Dynamic State Variable Models in Ecology: Methods and Applications.* Oxford, New York, 2000.

[35] F. H. Clarke. *Optimal Control and Nonsmooth Analysis.* SIAM, Philadelphia, 1990.

[36] D. Cohen. Maximizing final yield when growth is limited by time or by limiting resources. *Journal of Theoretical Biology,* 33:299–307, 1971.

[37] Michael M. Connors and Daniel Teichroew. *Optimal Control of Dynamic Operations Research Models.* International Textbook Company, Scranton, Pennsylvania, 1967.

[38] John B. Conway. *A Course in Functional Analysis.* Springer, New York, 1990.

[39] M. G. Crandall, L. C. Evans, and P.-L. Lions. Some properties of viscosity solutions for Hamiltonian-Jacobi equations. *Transactions American Mathematical Society,* 282, 1984.

[40] B. D. Craven. *Control and Optimization.* Chapman and Hall Mathematics, New York, 1995.

[41] J. M. Cushing. *Integrodifferential equations and delay models in population dynamics.* Springer-Verlag, New York, 1977.

[42] D. Daescu, G. R. Carmichael, and A. Sandu. Adjoint implementation of Rosenbrock methods applied to variational data assimilation problems. *Journal of Computational Physics,* 165:496–510, 2000.

[43] M. H. A. Davis. *Markov Models and Optimization.* Chapman and Hall, London, 1993.

[44] L. G. de Pillis, W. Gu, K. R. Fister, T. Head, K. Maples, T. Neal, A. Murugan, and K. Yoshida. Chemotherapy for tumors: An analysis of the dynamics and a study of quadratic and linear optimal controls. *Mathematical Biosciences,* to appear, 2007.

[45] D. L. DeAngelis, W. M. Post, and C. C. Travis. *Positive Feedback in Natural Systems.* Springer-Verlarg, Berlin, 1986.

[46] K. Renee Deaton. *Optimal Control of a Heat Flux in a Parabolic Partial Differential Equation.* Masters Thesis, University of Tennessee, 1992.

[47] Wandi Ding. Optimal control of a hybrid system and a fishery model. *University of Tennessee dissertation*, 2006.

[48] L. C. W. Dixson and M. C. Bartholomew-Biggs. Adjoint-control transformations for solving practical optimal control problems. *Optimal Control Applications and Methods*, 2:365–81, 1981.

[49] L. C. W. Dixson and M. C. Biggs. The advantage of adjoint-control transformations when determining optimal trajectories by Pontryagin's Maximum Principle. *Aeronautical Journal*, 76:169–74, 1972.

[50] A. L. Dontchev and W. W. Hager. The Euler approximation in state constrained optimal control. *Mathematics of Computation*, 70:173–303, 2001.

[51] Peter Dorato, Chaoki Abdallah, and Vito Cerone. *Linear Quadratic Control: An Introduction.* Prentice Hall, Englewood Cliffs, New Jersey, 2000.

[52] Z. Duda. Numerical solutions to bilinear models arising in cancer chemotherapy. *Nonlinear World*, 4(1):53–72, 1997.

[53] Leah Edelstein-Keshet. *Mathematical Models in Biology.* SIAM, Philadelphia, 2005.

[54] E. R. Edge and W. F. Powers. Function space quasi-Newton algoritms for optimal control problems having singular arcs. *Journal of Optimization Theory and Applications*, 20:455–79, 1976.

[55] Martin Eisen. *Mathematical Methods and Models in the Biological Sciences.* Prentice Hall, Englewood Cliffs, New Jersey, 1988.

[56] Lawrence C. Evans. *Partial Differential Equations.* American Mathematical Society, Providence, Rhode Island, 1998.

[57] H. O. Fattorini. *Infinite Dimensional Optimization and Control Theory.* Cambridge University Press, New York, 1999.

[58] A. F. Filippov. On certain questions in the theory of optimal control. *SIAM Journal on Control*, 1:76–84, 1968.

[59] K. Renee Fister. Optimal control of harvesting in a predator-prey parabolic system. *Houston J. of Math.*, 23:341–55, 1997.

[60] K. Renee Fister, S. Lenhart, and S. McNally. Optimizing chemotherapy in an HIV model. *Electronic Journal of Differential Equations*, 32:1–12, 1998.

[61] K. Renee Fister and John Carl Panetta. Optimal control applied to competing chemotherapeutic cell-kill strategies. *SIAM Journal of Applied Mathematics*, 63(6):1954–71, 2003.

[62] W. H. Fleming and R. W. Rishel. *Deterministic and Stochastic Optimal Control.* Springer-Verlag, New York, 1975.

[63] Irmgard Flugge-Lotz. *Discontinuous and Optimal Control.* McGraw-Hill, New York, 1969.

[64] G. Fraser-Andrews. Numerical methods for singular optimal control. *Journal of Optimization Theory and Applications*, 61(3):377–401, 1989.

[65] A. Friedman. *Stochastic Differential Equations and Applications.* Academic Press, New York, 1975.

[66] Avner Friedman. *Partial Differential Equations.* Holt, Rinehart, and Winston, New York, 1969.

[67] H. Gaff, H. R. Joshi, and S. Lenhart. Optimal harvesting during an invasion of a sublethal plant pathogen. *Environment and Development Economics*, to appear 2007.

[68] Wayne M. Getz. Optimal control and principles in population management. *Proceedings of Symposia in Applied Mathematics*, 30:63–82, 1984.

[69] Mini Ghosh and Andrea Pugliese. Seasonal population dynamics of ticks, and its influence on infection transmission: a semi-discrete approach. *Bulletin of Mathmetical Biology*, 66:1659–84, 2004.

[70] Bean San Goh, George Leitmann, and Thomas L. Vincent. Optimal control of a prey-predator system. *Mathematical Biosciences*, 19, 1974.

[71] John Gregory and Cantian Lin. *Constrained Optimization in the Calculus of Variations and Optimal Control.* Van Nostrand Reinhold, New York, 1992.

[72] A. K. Griffith and N. K. Nichols. Adjoint techniques in data assimilation for treating systematic model error. *Journal of Flow, Turbulence and Combustion*, 65:469–88, 2001.

[73] V. Grimm and S. F. Railsback. *Individual-Based Modeling and Ecology.* Princeton University Press, Princeton, New Jersey, 2005.

[74] W. Hackbush. A numerical method for solving parabolic equations with opposite orientations. *Computing*, 20(3):229–40, 1978.

[75] Wassim M. Haddad, Vijay Sekhar Chellaboina, and Sergey G. Nersesov. *Impulsive and Hybrid Dynamical Systems.* Princeton University Press, New York, 2006.

[76] J. K. Hale. *Theory of Functional Differential Equations.* Springer-Verlag, New York, 1977.

[77] J. K. Hale and S. N. Verduyn Lunel. Stability and control of feedback systems with time delays. *International Journal of Systems Science*, 34:497–504, 2003.

[78] H. Halkin. A maximum principle of the Pontryagin type for systems described by nonlinear differential equations. *J. SIAM Control*, 4:90–111, 1966.

[79] I. Hanski. *Metapopulation Ecology.* Oxford University Press, Oxford, 1999.

[80] R. F. Hartl, S. P. Sethi, and R. G. Vickson. A survey of the maximum principles for optimal control problems with state constraints. *SIAM Review*, 37, 1995.

[81] Bradford A. Hawkins and Howard V. Cornell editors. *Theoretical Approaches to Biological Control.* Cambridge, Cambridge, UK, 1999.

[82] A. Heinricher, S. Lenhart, and A. Solomon. The application of optimal control methodology to a well-stirred bioreactor. *Natural Resource Modeling*, 9:61–80, 1995.

[83] G. E. Herrera. The benefits of spatial regulation in a multispecies fishery. *Marine Resource Economics*, 21:63–79, 2006.

[84] John Hof and Michael Bevers. *Spatial Optimization for Managed Ecosystems.* Columbia University Press, New York, 1998.

[85] John Hof and Michael Bevers. *Spatial Optimization in Ecological Applications.* Columbia University Press, New York, 2002.

[86] A. I. Houston and J. M. McNamara. *Models of adaptive behavior: An approach based on state.* Cambridge University Press, Cambridge, 1999.

[87] Michael D. Intriligator. *Mathematical Optimization and Economic Theory.* SIAM, Philadelphia, 2002.

[88] K. Ito. On stochastic differential equations. *Memoirs of American Mathematical Society*, 4:1–51, 1951.

[89] D. S. Jones and B. D. Sleeman. *Differential Equations and Mathematical Biology.* Chapman and Hall/CRC Press, Boca Raton, Florida, 2003.

[90] H. R. Joshi. Optimal control of an HIV immunology model. *Optimal Control Applic. Methods*, 23:199–213, 2003.

[91] H. R. Joshi, S. Lenhart, and H. Gaff. Optimal harvesting in an integrodifference population model. *Optimal Control Applications and Methods*, 27:61–75, 2006.

[92] H. R. Joshi, S. Lenhart, T. Herrera, and M. G. Neubert. Optimal dynamic harvest of a diffusive renewable resource. *Preprint*, 2007.

[93] H. R. Joshi, S. Lenhart, M. Y. Li, and L. Wang. Optimal control methods applied to disease models. *AMS Volume on Mathematical Studies on Human Disease Dynamics Emerging Paradigms and Challenges*, 410:187–207, 2006.

[94] H. R. Joshi, S. Lenhart, H. Lou, and H. Gaff. Harvesting control in an integrodifference population model with concave growth term. *Hybrid Systems and Applications*, to appear, 2007.

[95] E. Jung, S. Lenhart, and Z. Feng. Optimal control of treatments in a two strain tuberculosis model. *Discrete and Continuous Dynamical Systems*, 2:473–82, 2002.

[96] E. Jung, S. Lenhart, V. Protopopescu, and C. Babbs. Optimal control theory applied to a difference equation model for cardiopulmonary resucitation. *Mathematical Models and Methods in Applied Sciences*, 15:307–21, 2005.

[97] E. Jung, S. Lenhart, V. Protopopescu, and C. Babbs. Optimal strategy for cardiopulmonary resucitation with continuous chest compression. *Academic Emergency Medicine Journal*, 13:715–22, 2006.

[98] E. Jung, S. Lenhart, V. Protopopescu, and Y. Braiman. Optimal control of transient behavior coupled solid state lasers. *Physical Review B*, 67:046222–6, 2003.

[99] Velimir Jurdjevic. *Geometric Control Theory*. Cambridge University Press, New York, 1997.

[100] Morton I. Kamien and Nancy L. Schwartz. *Dynamic Optimization: The Calculus of Variations and Optimal Control in Economics and Management*. North-Holland, New York, 1991.

[101] Y. H. Kang, S. Lenhart, and V. Protopopescu. Optimal control of parameter and input functions for nonlinear systems. *Houston J. of Mathematics*, to appear, 2007.

[102] C. T. Kelley. *Iterative Methods for Optimization*. SIAM, Philadelphia, 1999.

[103] D. Kinderlehrer and G. Stampachia. *An Introduction to Variational Inequalities and Their Applications*. Academic Press, new edition SIAM, 2000, New York, 1980.

[104] David King and Jonathan Roughgarden. Graded allocation between vegetative reproductive growth for annual plants in growing seasons of random length. *Theoretical Population Biology*, 22(1):1–16, 1982.

[105] D. Kirschner and G. F. Webb. A model for treatment strategy in the chemotherapy of AIDS. *Bulletin of Mathematical Biology*, 58(2):367–92, 1996.

[106] Greg Knowles. *An Introduction to Applied Optimal Control*. Academic Press, New York, 1981.

[107] Mark Kot. *Elements of Mathematical Ecology*. Cambridge Press, Cambridge, 2001.

[108] Mark Kot. Do invading organisms do the wave? *Canadian Applied Mathematics Quarterly*, 10:139–70, 2002.

[109] A. J. Krener. The high order maximum principle and its application to singular extremals. *SIAM Journal on Control and Optimization*, 15:256–93, 1977.

[110] Irena Lasiecka. *Mathematical Control Theory of Coupled PDEs*. SIAM, Philadelphia, Pennsylvania, 2002.

[111] Irena Lasiecka and Roberto Triggiani. *Control Theory for Partial Differential Equations: Continuous and Approximation Theories: I. Abstract Parabolic Systems*. Cambridge University Press, New York, 2000.

[112] Irena Lasiecka and Roberto Triggiani. *Control Theory for Partial Differential Equations: Continuous and Approximation Theories: II. Abstract Hyperbolic-like Systems over a Finite Time Horizon*. Cambridge University Press, New York, 2000.

[113] U. Ledzewicz, T. Brown, and H. Schattler. A comparison of optimal controls for a model in cancer chemotherapy with L_1- and L_2-type objectives. *Optimization Methods and Software*, 19:351–9, 2004.

[114] U. Ledzewicz and H. Schaettler. Second order conditions for extremum problems with nonregular equality constraints. *Journal of Optimization, Theory and Applications*, 86:113–44, 1995.

[115] U. Ledzewicz and H. Schattler. Optimal control for a bilinear model with recruiting agent in cancer chemotherapy. *Proceedings of the 42nd IEEE Conference on Decision and Control*, pages 2762–2767, 2003.

[116] E. B. Lee and L. Markus. *Foundations of Optimal Control Theory*. Wiley, New York, 1967.

[117] S. Lenhart and M. Bhat. Application of distributed parameter control model in wildlife damage management. *Mathematical Models and Methods in Applied Sciences*, 2:423–39, 1993.

[118] S. Lenhart and M. Liang. Bilinear optimal control for a wave equation with viscous damping. *Houston Journal of Mathematics*, 26:575–95, 2000.

[119] S. Lenhart, M. Liang, and V. Protopopescu. Optimal control of the effects of the boundary habitat hostility. *Math. Methods in Applied Sciences*, 22, 1999.

[120] S. Lenhart, V. Protopopescu, and J. T. Workman. Minimizing transient time in a coupled solid-state laser model. *Mathematical Methods in Applied Sciences*, 29:373–86, 2006.

[121] S. Lenhart, T. Seidman, and J. Yong. Optimal control of a bioreactor with modal switching. *Mathematical Models and Methods in Applied Sciences*, 11:933–49, 2001.

[122] S. Lenhart, S. Stojanovic, and V. Protopopescu. A minimax problem for semilinear nonlocal competitive systems. *Applied Math and Optimization*, 28:113–132, 1993.

[123] S. Lenhart and D. G. Wilson. Optimal control of a heat transfer problem with convective boundary conditions. *J. Optimization Theory and Applications*, 79:581–97, 1993.

[124] D. Leonard and N. V. Long. *Optimal Control Theory and Static Optimization in Economics*. Cambridge University Press, Cambridge, 1998.

[125] A. Leung and S. Stojanovic. Optimal control for elliptic volterra-lotka type equations. *J. Math. Anal. Applications*, 173:603–19, 1993.

[126] Frank L. Lewis. *Optimal Control*. Wiley-Interscience, New York, 1986.

[127] M. A. Lewis and R. W. Van Kirk. Integrodifference models for persistence in fragmented habitats. *Bulletin of Mathematical Biology*, 49:10–137, 1977.

[128] Xunjing Li and Jiongmnin Yong. *Optimal Control Theory for Infinite Dimensional Systems*. Birkhäuser, Boston, 1995.

[129] J. L. Lions. *Optimal Control of Systems Governed by Partial Differential Equations*. Springer-Verlag, Berlin, 1971.

[130] Arturo Locatelli. *Optimal Control*. Birkhäuser, Boston, 2001.

[131] D. L. Lukes. *Differential Equations: Classical to Controlled*. Academic Press, New York, 1982.

[132] Rein Luus. Application of dynamic programming to singular optimal control problems. *Proceedings of the 1990 American Control Conference*, pages 2932–7, 1990.

[133] Rein Luus. On the application of iterative dynamic programming to singular optimal control problems. *IEEE Transactions on Automatic Control*, 37(11):1802–6, 1992.

[134] Jim Ma and Jiongmin Yong. *Forward-Backward Stochastic Differential Equations and Their Applications*. Springer, New York, 1999.

[135] J. Macki and A. Strauss. *Introduction to Optimal Control Theory*. Springer-Verlag, New York, 1982.

[136] Marc Mangel and C. W. Clark. *Dynamics Modeling in Behavorial Ecology*. Princeton University Press, New Jersey, 1988.

[137] C. Martin, L. Allen, M. Stamp, M. Jones, and R. Carpio. A model for the optimal control of measles. *Computational and Control II: Proceedings of the Third Bozeman Conference*, 3:265–83, 1992.

[138] Matlab. *Matlab Release 13*. The Mathworks, Inc., Natich, MA, 2003.

[139] R. Mattheij and J. Molnaar. *Ordinary Differential Equations in Theory and Practice*. Wiley, Chichester, UK, 1996.

[140] H. Maurer. Numerical solutions of singular control problems using multiple shooting techniques. *Journal of Optimization Theory and Applications*, 18:235–57, 1976.

[141] Ian McCausland. *Introduction to Optimal Control*. John Wiley & Sons, New York, 1969.

[142] J. P. McDanell and W. F. Powers. Necessary conditions for joining optimal singular and nonsingular subarcs. *SIAM Journal on Control*, 9:161–73, 1971.

[143] N. McDonald. *Time lags in biological models*. Springer-Verlag, Berlin, 1978.

[144] Boris M. Miller and Evgeny Ya. Rubinovich. *Impulsive Control in Continuous and Discrete-Continuous Systems*. Kluwer Academic/Plenum Publishers, New York, 2003.

[145] M. E. Moody and R. N. Mack. Controlling the spread of plant invasions: the importance of nascent foci. *Journal of Applied Ecology*, 25:1009–21, 1988.

[146] Douglas Mooney and Randall Swift. *A Course in Mathematical Modeling*. Mathematical Association of American, United States, 1999.

[147] Boris S. Mordukhovich. *Variational Analysis and Generalized Differentiation I, II*. Springer, Berlin, 2006.

[148] Kirsten Morris. *Introduction to Feedback Control*. Harcourt, San Diego, California, 2001.

[149] R. Morton and K. H. Wickwire. On the optimal control of a deterministic epidemic. *Advances in Applied Probability*, 6:522–635, 1974.

[150] J. D. Murray. *Mathematical Biology II: Spatial Models and Biomedical Applications*. Springer-Verlag, New York, 2003.

References

[151] R. M. Murray, Z. Li, and S. S. Sastry. *A Mathematical Introduction to Robotic Manipulation.* CRC Press, Boca Raton, 1994.

[152] P. Neittanmaki and D. Tiba. *Optimal Control of Nonlinear Parabolic Systems: Theory, Algorithms, and Applications.* Marcel Dekker, New York, 1994.

[153] M. G. Neubert, M. Kot, and M. A. Lewis. Dispersal and pattern formation in a discrete-time predator-prey model. *Theoretical Population Biology*, 48:7–43, 1995.

[154] Michael G. Neubert. Marine reserves and optimal harvesting. *Ecology Letters*, 6:843–9, 2003.

[155] C. Neuhauser. Mathematical challenges in spatial ecology. *AMS Notices*, 48:1304–14, 2004.

[156] D. A. Nomirovs'kyi. On the convergence of gradient methods used in solving singular optimal control problems. *Journal of Mathematical Sciences*, 107(2):3787–92, 2001.

[157] Peter Ögren and Clyde F. Martin. Vaccination strategies for epidemics in highly mobile populations. *Applied Mathematics and Computation*, 127:261–76, 2002.

[158] L. S. Pontryagin, V. G. Boltyanskii, R. V. Gamkrelize, and E. F. Mishchenko. *The Mathematical Theory of Optimal Processes.* Wiley, New York, 1962.

[159] Graham F. Raggett. An efficient gradient technique for the solution of optimal control problems. *Computer Methods in Applied Mechanics and Engineering*, 12:315–22, 1977.

[160] T. J. Regan, M. A. McCarthy, P. W. J. Baxter, F. D. Panetta, and H. P. Possingham. Optimal eradication: when to stop looking for an invasive plant. *Ecology Letters*, 9:759–66, 2006.

[161] Sheldon M. Ross. *Introduction to Stochastic Dynamic Programming.* Academic Press, New York, 1983.

[162] H. L. Royden. *Real Analysis.* McMillan Publishing Company, Inc., New York, 1968.

[163] Walter Rudin. *Real and Complex Analysis.* McGraw-Hill Book Company, New York, 1987.

[164] D. L. Russell. *Mathematics of finite-dimensional control systems, theory and design.* Marcel Dekker, New York, 1979.

[165] R. A. Salinas, S. Lenhart, and L. J. Gross. Control of a metapopulation harvesting model for black bears. *Natural Resource Modeling*, 18:307–21, 2005.

[166] J. N. Sanchirico, U. Malvadkar, A. Hasting, and J. E. Wilen. When are no-take zones an economically optimal fishery management strategy? *Ecological Applications*, 16:1643–59, 2006.

[167] Atle Seierstad and Knut Sydaester. *Optimal Control Theory with Economic Applications*. North Holland, Amsterdam, 1987.

[168] S. Ya. Serovaiskii. *Counterexamples in Optimal Control Theory*. VSP, Brill Academic Publishers, Utrecht, 2004.

[169] S. P. Sethi and G. L. Thompson. *Optimal Control Theory; Applications to Management Science and Economics*. Kluwer, Boston, 2000.

[170] H. Smith and P. Waltman. *The theory of the chemostat*. Cambridge University Press, Cambridge, 1995.

[171] Elias Stein and Rami Shakarchi. *Real Analysis: Measure Theory, Integration, and Hilbert Spaces*. Princeton University Press, Princeton, New Jersey, 2005.

[172] H. J. Sussman and V. Jurdjevic. Controllability of nonlinear systems. *Journal of Differential Equations*, 12, 1972.

[173] G. W. Swan. *Applications of Optimal Control Theory in Biomedicine*. Marcel Dekker, New York, 1987.

[174] K. C. Tan and R. J. Bennett. *Optimal Control of Spatial Systems*. George Allen and Unwin, London, 1984.

[175] K. L. Teo, C. J. Goh, and K. H. Wong. *A Unified Computational Approach to Optimal Control Problems*. John Wiley & Sons, New York, 1991.

[176] Fredi Tröltzsch. *Optimality Condtions for Parabolic Control Problems and Applications*. Teubner, Leipzig, Germany, 1984.

[177] William R. Wade. *An Introduction to Analysis*. Prentice Hall, Upper Saddle River, New Jersey, 2000.

[178] A. Whittle, S. Lenhart, and J. White. Optimal control of gypsy moth populations. *Preprint*, 2007.

[179] A. J. Whittle, S. Lenhart, and L. J. Gross. Optimal control for management of an invasive plant species. *Mathematical Biosciences and Engineering*, to appear, 2007.

[180] K. H. Wickwire. Mathematical models for the control of pests and infectious diseases: A survey. *Theoretical Population Biology*, 11:182–238, 1977.

[181] J. Wu. *Theory and Applications of Partial Functional Differential Equations*. Springer, New York, 1996.

[182] Jiongmin Yong and Xun Yu Zhou. *Stochastic Controls: Hamiltonian Systems and HJB Equations*. Springer-Verlag, New York, 1999.

[183] M.I. Zelikin and V. F. Borisov. *Theory of Chattering Control with Applications to Astronautics, Robotics, Economics, and Engineering*. Birkhäuser, Boston, 1994.

Index

adapted forward-backward sweep method, 178, 190, 209
adjoint
    discrete, 194
    equation, 11, 12, 38, 98
    operator, 214, 222
    variable, 8, 10, 26, 38, 73
admissible, 42
Allee effect, 19, 115
annual plants, 1
autonomous, 33

bacteria, 67
bang-bang control, 140, 154, 159, 171
Beverton-Holt stock-recruitment model, 193
bioreactor, 102, 157, 187
bisection method, 182
bounds on the control, 71, 85
bounds on the state, 242

$CD4^+T$ cell, 123
characterization of the optimal control, 15, 50, 81
chattering controls, 243
chemotherapy, 89, 123
co-metabolism, 102, 157
concave, 5
continuously differentiable, 5, 7, 163
control, 3, 7
    discrete, 193
controllability, 240
convex, 5, 215
convex combination, 50, 56, 82, 141, 156, 206
CPR, 242

data assimilation, 243

degenerate, 42
delays, 241
diabetes, 135
discount term, 153, 216, 221
discrete time models, 193
dynamic programming, 237
    stochastic, 239
dynamic state variable models, 238

eliminating adjoint variables, 104
eliminating state variables, 101–104, 118, 158
elliptic, 216
epidemic disease, 117
equilibrium, 63
existence, 25, 212

feedback control, 105, 240
fishery, 93
flux control, 232
forward-backward sweep method, 50, 57, 175, 190
    adapted method, 178, 190, 209
    bang-bang controls, 140
    bounds on the control, 81–82
    discrete, 198
    discrete case, 206
    multiple states and controls, 112
    pde, 215, 220, 223
free terminal time, 163, 183
fungicide, 63

gain matrix, 106
geometric control theory, 240
glucose, 135
glucose tolerance test, 135
Gompertzian growth, 89
gradient, 243

260   Index

Hamilton-Jacobi-Bellman
    equation, 238
Hamiltonian, 12, 31, 75, 98, 105, 139, 165
    as a constant function, 34
    as a Lipschitz function, 32
    discrete, 196
harvesting model, 115, 216
    black bear, 129–133
    fish, 71, 93–95, 146, 149–150
    timber, 153–156
HIV, 123
hybrid systems, 241

incidence, 117
incubation period, 117
infinite time horizon problems, 240
insulin, 135
integrodifferential equations, 241
invasive plant species, 201, 205
isoperimetric constraint, 109, 115, 181, 189
Ito calculus, 241

jumps, 241

Laplacian, 216
Lebesgue integrable, 25, 212
Legendre-Clebsh condition, 148
linear programming, 239
linear quadratic regulator, 104, 114, 233
Lipschitz, 6, 26
    Hamiltonian, 32
log-kill hypothesis, 89
Lotka-Volterra model, 189, 223, 228

MATLAB, 52, 57
    largest number, 66, 68
Mean Value Theorem, 7
metabolism, 157
metapopulation, 129
Michealis-Menton kinetics, 102
minimal time problem, *see* time optimal control

mold, 63
Monod kinetics, 102
multiple controls, 97
    discrete, 199
    with bounds, 98
multiple states, 97
    discrete, 199

necessary conditions, 8, 12, 25, 37, 47, 75, 83, 165, 196
Neumann boundary conditions, 223, 228, 231
Newton's method, 175
no-flux, 150
no-take marine reserve, 149, 220
nonlinear programming, 239

objective functional, 3, 7
    discrete, 194
observability, 240
optimal control, 7
optimality condition, 12, 38, 98
    discrete, 196
optimality system, 15, 214
overdetermined, 42

parabolic, 216, 223, 231
parameter identification, 228, 242
payoff term, 37, 67, 77, 97, 182, 194
pesticide, 189
photosynthate, 1
piecewise continuous, 4, 243
piecewise differentiable, 5
Pontryagin's Maximum Principle, 12, 31, 75
predator-prey model, 189–192, 223, 228
present-value, 153
principle of optimality, 28, 35, 237
protease inhibitors, 123

relative error, 51
reverse transcription inhibitors, 123
Riccati equation, 105
round-off error, 155

Runge-Kutta, 50, 57, 198
  for systems, 112

salvage term, *see* payoff term
satellite, 201, 205
secant method, 175, 190
second order condition, 148
SEIR model, 117
sensitivity, 194, 213
shadow price, 27
shooting method, 49, 178
singular control, 140, 143
Sobolev space, 211
stabilization, 240
state variable, 3, 7
  discrete, 193
  fixed endpoints, 42, 75, 98, 165, 177, 184
steady state equation, 149
stochastic differential equations, 241
sufficient conditions, 8, 23
sweep method, 105
switching function, 139
switching times, 140, 156, 159

tangent line property, 5, 14, 24
Tikhonov's regularization, 229
time optimal control, 168
transversality condition, 11, 12, 38, 42, 75, 98, 195, 205
truncation, 78–80, 82, 85, 86
tumor cells, 40, 89, 109, 163

uniqueness, 26, 215

vaccination, 117–122
value function, 26
variational inequalities, 242
viscosity solutions, 238

weak derivative, 211
weight parameter, 61, 63, 85, 93, 120, 158

CPSIA information can be obtained
at www.ICGtesting.com
Printed in the USA
LVHW040558150219
607587LV00001B/1/P